NUTRIENT
REQUIREMENTS
OF
DOMESTIC
ANIMALS

NUMBER 15

Nutrient Requirements of Goats:

Angora, Dairy, and Meat Goats in Temperate and Tropical Countries

Subcommittee on Goat Nutrition
Committee on Animal Nutrition
Board on Agriculture and Renewable Resources
Commission on Natural Resources
National Research Council

NATIONAL ACADEMY PRESS
Washington, D.C. 1981

NUTRIENT REQUIREMENTS OF DOMESTIC ANIMALS SERIES

Nutrient Requirements of Mink and Foxes, 1968	0-309-01676-2
Nutrient Requirements of Trout, Salmon, and Catfish, 1973	0-309-02141-3
Nutrient Requirements of Dogs, 1974	0-309-02315-7
Nutrient Requirements of Sheep, 1975	0-309-02212-6
Nutrient Requirements of Beef Cattle, 1976	0-309-02419-6
Nutrient Requirements of Rabbits, 1977	0-309-02607-5
Nutrient Requirements of Warmwater Fishes, 1977	0-309-02616-4
Nutrient Requirements of Poultry, 1977	0-309-02725-X
Nutrient Requirements of Cats, 1978	0-309-02743-8
Nutrient Requirements of Horses, 1978	0-309-02760-8
Nutrient Requirements of Dairy Cattle, 1978	0-309-02749-7
Nutrient Requirements of Nonhuman Primates, 1978	0-309-02786-1
Nutrient Requirements of Laboratory Animals, 1978	0-309-02767-5
Nutrient Requirements of Swine, 1979	0-309-02870-1
Nutrient Requirements of Coldwater Fishes, 1981	0-309-03187-7

NOTICE: The project that is the subject of this report was approved by the Governing Board of the National Research Council, whose members are drawn from the councils of the National Academy of Sciences, the National Academy of Engineering, and the Institute of Medicine. The members of the committee responsible for the report were chosen for their special competences and with regard for appropriate balance.

This report has been reviewed by a group other than the authors according to procedures approved by a Report Review Committee consisting of members of the National Academy of Sciences, the National Academy of Engineering, and the Institute of Medicine.

The National Research Council was established by the National Academy of Sciences in 1916 to associate the broad community of science and technology with the Academy's purposes of furthering knowledge and of advising the federal government. The Council operates in accordance with general policies determined by the Academy under the authority of its congressional charter of 1863, which establishes the Academy as a private, nonprofit, self-governing membership corporation. The Council has become the principal operating agency of both the National Academy of Sciences and the National Academy of Engineering in the conduct of their services to the government, the public, and the scientific and engineering communities. It is administered jointly by both Academies and the Institute of Medicine. The National Academy of Engineering and the Institute of Medicine were established in 1964 and 1970, respectively, under the charter of the National Academy of Sciences.

This study was supported by the Agency for International Development, U.S. Department of State, the Science and Education Administration of the U.S. Department of Agriculture, and the Bureau of Veterinary Medicine, Food and Drug Administration of the U.S. Department of Health and Human Services. Support was also received from the American Dairy Goat Association and the Caprine Research Foundation.

Library of Congress Catalog Card Number: 81-84592

International Standard Book Number: 0-309-03185-0

First Printing, December 1981
Second Printing, September 1989
Third Printing, November 1992

Available from

NATIONAL ACADEMY PRESS
2101 Constitution Avenue, N.W.
Washington, D.C. 20418

Printed in the United States of America

PREFACE

This report is one of a series issued under the direction of the Committee on Animal Nutrition, Board on Agriculture and Renewable Resources, National Research Council (NRC). It was prepared by the Subcommittee on Goat Nutrition. Its use may be enhanced by referring also to three other reports in the series on Nutrient Requirements of Domestic Animals: Number 3, *Nutrient Requirements of Dairy Cattle,* fifth revised edition, 1978; Number 5, *Nutrient Requirements of Sheep,* fifth revised edition, 1975; and Number 6, *Nutrient Requirements of Horses,* fourth revised edition, 1978.

The first two parts of Chapter 1, on energy and protein requirements, contain much new information from extensive research at the Raja Balwant Singh College at Bichpuri (Agra), India, under support from USDA PL-480 funding for dairy and meat goats; and from research at the Texas A&M Agricultural Experiment Station at San Angelo for Angora goats. The final three parts of Chapter 1, on mineral, vitamin, and water requirements, and Chapter 5 on nutrition-related metabolic disorders have had to rely mostly on reviewing past work. Chapter 3 on herbage and browse utilization includes new work from Texas. Chapter 4 on ration formulation and examples of typical rations, and Chapter 2 with tables of nutrient requirements, involve extensive, previously unpublished data resources and worldwide experiences of each subcommittee member. Because of the unique nature of goats, the nutrient requirement tables also include allowances for activities during grazing. The feed composition tables also contain new data from research by the subcommittee members and significant contributions by the International Feedstuffs Institute at Utah State University. The inclusion of an extensive number of references in the bibliography was considered desirable for the benefit of future studies, although not all were cited directly in the text.

The subcommittee is indebted to Philip Ross and Selma P. Baron of the Board on Agriculture and Renewable Resources for their assistance in the production of this report and to the members of the Committee on Animal Nutrition for their valuable suggestions and reviews. Special thanks are due A. N. Bhattacharya (Turkey), Tony J. Cunha (California), R. E. McDowell (New York), Pierre Morand-Fehr (France), Amiram Shkolnik (Israel), and Jim Yazman (Arkansas) for their comprehensive reviews and constructive comments, and Bernard S. Schweigert and colleagues (University of California, Davis) and Glenn W. Salisbury (Illinois) who reviewed the report for the Board on Agriculture and Renewable Resources and Commission on National Resources, respectively. The subcommittee also expresses appreciation to Ivan Lindahl for his pioneering efforts in the initiation of this subcommittee.

Subcommittee on Goat Nutrition

GEORGE F. W. HAENLEIN, *Chairman*; University of Delaware, Newark
CANAGASABY DEVENDRA, Malaysian Agricultural Research and Development Institute, Serdang, Malaysia
JAMES E. HUSTON, Texas A&M University, San Angelo
O. P. S. SENGAR, Raja Balwant Singh College, Bichpuri (Agra), India
MAURICE SHELTON, Texas A&M University, San Angelo
S. N. SINGH, Raja Balwant Singh College, Bichpuri (Agra), India

Committee on Animal Nutrition

JOSEPH P. FONTENOT, *Chairman*; Virginia Polytechnic Institute and State University
CARL E. COPPOCK, Texas A&M University
RICHARD D. GOODRICH, University of Minnesota
BERYL E. MARCH, University of British Columbia
PAUL W. MOE, USDA Animal Science Institute
WILSON G. POND, USDA Meat Animal Research Center
QUINTON R. ROGERS, University of California, Davis
GARY L. RUMSEY, Tunison Laboratory of Fish Nutrition
DUANE E. ULLREY, Michigan State University

SELMA P. BARON, *Staff Officer*

CONTENTS

1 INTRODUCTION

This report represents a first effort to establish the nutrient requirements of goats from original studies directly concerned with the needs of goats. Current scientific literature on goats is not nearly as extensive and comprehensive as it is for other domestic livestock species, and past efforts to set nutrient requirements have relied heavily on extrapolation of values derived from cattle and sheep studies.

There are approximately 400 million goats in the world, with over 2 million head in the United States alone, producing meat, milk, fiber (mohair) and skins (Haenlein, 1978; FAO, 1979; Shelton, 1980a,b; Terrill, 1980). An increasing number of people in the United States and around the world drink milk from goats, and interest in goats is on the increase (Bhattacharya, 1980; Devendra, 1980a; Haenlein, 1980a; Harris, 1980; Iloeje *et al.*, 1980; Leach, 1980). There is an increasing place for goats in commercial agriculture and with the small, part-time farmer (Lowe, 1943; Woinoff, 1949; Shannon, 1956; Garcia, 1958; Clamohoy *et al.*, 1959; Agricultural Research Institute, Cyprus, 1964; Oppong, 1965; Wahid, 1965; Banco Nacional Agropecuario, 1971; Skinner, 1972; Anonymous, 1973; Choveiri, 1973; Joubert, 1973; Maubecin, 1973; Bula *et al.*, 1977; McDowell and Bove, 1977; Ace, 1978; Fitzhugh *et al.*, 1978; Guss, 1978; Sands and McDowell, 1978).

Several national and international goat symposia have been staged recently: the American Dairy Science Association symposium in 1977 at Iowa State University, and in 1979 at Utah State University; the French National Institute for Sheep and Goat Research (ITOVIC) symposia in 1971 and 1981 at Tours; and the Winrock International Livestock Research and Training Center symposia in 1977 and 1978 at Morrilton, Arkansas. The Third International Conference on Goat Production and Disease will meet in 1982 at Tucson, Arizona. Their proceedings add greatly to the knowledge of the nutrient requirements of goats.

It is also increasingly evident that despite similarities to sheep and cattle, goats exhibit significant differences in grazing habits, physical activities, water requirements, feed selection, milk composition, carcass composition, metabolic disorders, and parasites. The nutrient requirements of goats may therefore justifiably be treated separately from those of other ruminants. As older literature becomes more accessible and new communication among world scientists increases, further documentation and information will be forthcoming (Nottbohm and Phillipi, 1933; Zorn *et al.*, 1938; Meklenburcev, 1949; Benzie and Phillipson, 1957; Saperstein, 1960; Crawford and Grogan, 1961; Eker, 1961; Hanson and Andersen, 1962; Hofmeyr *et al.*, 1965; Schaefer, 1967; Rogers *et al.*, 1969; Brandsch and Kruger, 1970; French, 1970; Morand-Fehr and deSimiane, 1977; Corbett, 1978; Haenlein, 1980a,b; Jenness, 1980). This first NRC report on the nutrient requirements of goats must be considered within the limits of available knowledge, and refinements are reserved for subsequent editions as the literature of goats improves.

2 NUTRIENT REQUIREMENTS

ENERGY

Efficient utilization of nutrients depends on an adequate supply of energy, which is of paramount importance in determining the productivity of goats. Energy deficiency retards kid growth, delays puberty, reduces fertility, and depresses milk production (Singh and Sengar, 1970; Sachdeva *et al.*, 1973). With continued deficiency the animals show a concurrent reduction in resistance to infectious diseases and parasites. The problem may be further complicated by deficiencies of protein, minerals, and vitamins.

Energy limitations may result from inadequate feed intake or from the low quality of the diet. Low energy intake that results from either feed restriction or low diet component digestibility prevents goats from meeting their requirements and from attaining their genetic potential. High water content of forages may also become a limiting factor.

Energy requirements are affected by age, body size, growth, pregnancy, and lactation, which have been treated as separate items in presenting requirements (Table 1). Energy requirements are also affected by the environment, hair growth, muscular activity, and relationships with other nutrients in the diet which, for best results, need to be supplied in adequate amounts. Temperature, humidity, sunshine, and wind velocity may increase or decrease energy needs depending upon the region. Stress of any kind may increase energy requirements.

Shearing mohair from Angora goats and pashmina from Cashmere goats decreases insulation and results in increased energy needs, especially during cold weather. Goats are more active and travel greater distances than sheep, which increases energy requirements. Maintenance requirements of goats on pasture, browse, and range, especially in mountainous and transhumance grazing, are considerably higher than those of stable-fed animals. The magnitude of this increase is considered in Table 1 at three levels of activity, and depends on availability of feed, water, topography, elevation and distances travelled in grazing.

Good quality roughages furnish about 2 Mcal metabolizable energy (ME) per kg dry matter (DM). Roughage-concentrate mixed rations are sometimes necessary to increase the energy content of the diet to 2.5 or 3.0 Mcal ME/kg DM when feeding early weaned kids or high-producing dairy goats. The efficiency with which energy is utilized for weight gain, pregnancy, and lactation usually increases with increasing levels of ME concentration in the diet.

It is probable that under certain conditions goats require a minimum of fats in their diet (Fehr and Delage, 1973; Morand-Fehr and Sauvant, 1980), but more studies are needed to better define these requirements.

The energy requirements of goats reported here are amounts needed for maintenance (including physical activity), optimum growth, reproduction, and milk and fiber production; they do not represent maximum levels under *ad libitum* intake. Energy requirements for the various categories of goats in Table 1 have been expressed as digestible energy (DE), ME, and net energy (NE) for maintenance, gain, pregnancy, lactation, and fiber production. Figures for total digestible nutrients (TDN) have been included because they are still widely in use around the world. For converting one form of energy into another the recommendations of Garrett *et al.* (1959) have been used: 100 Mcal gross energy (GE) = 76 Mcal DE = 62 Mcal ME = 35 Mcal NE. These values are higher than would be expected in forage-only diets; however, their relative relationships would still hold. For converting one standard of energy requirement into another, the following equalities have been employed (two or three significant decimal figures are included to reduce rounding errors):

1 kg starch equivalent (SE) or European starch unit (ESU) = 5.082 Mcal DE = 2.356 kcal NE
1 kg digestible organic matter (DOM) = 1.05 kg TDN
1 kg SE or 1 ESU = 1.15 kg TDN = 1.10 kg DOM
1 kg DOM = 4.620 Mcal DE
1 kg TDN = 4.409 Mcal DE
1 feed fattening unit (ffu) = 1.650 Mcal ME

1 European feed unit (Efu) = 2.500 kcal NE
1 kilojoule (kJ) = 0.239 kcal
1 kcal = 4.184 kJ

Maintenance

Energy requirements for maintenance of goats have been derived from pooled means of experimental data reported in terms of kcal ME/$W_{kg}^{0.75}$ per day. These include: 111.00 (Haenlein, 1950); 115.09 (Majumdar, 1960); 90.35 (Devendra, 1967a); 109.93 (Singh and Sengar, 1970); 100.00 (Flatt *et al.*, 1972); 92.92 (Akinsoyinu, 1974); 91.87 (Winter and Goersch, 1974); 87.31 (Itoh *et al.*, 1979); 101.98 (Rajpoot, 1979); 113.34 (Sengar, 1980). The average is 101.38 kcal ME/$W_{kg}^{0.75}$, which is comparable to figures provided for sheep: 98.31 (ARC, 1965); 95.75 (Olatungi, 1974); 91.99 (Adu, 1975); and 97.58 (NRC, 1975). But the figure is low compared to those for dairy cattle: 128.58 (ARC, 1965); 113.02 (NRC, 1978); and 115.22 (Rattray *et al.*, 1974).

The mean value of 101.38 kcal ME/$W_{kg}^{0.75}$ has been used in Table 1 to determine goat maintenance requirements for body weights ranging from 10 to 100 kg. The energy requirements have also been expressed as they vary with energy concentration in the diet from 2.0 to 2.4 Mcal ME/kg DM.

Activity

The basic ME requirement has also been used to calculate values for grazing at three levels of muscular activity. A 25 percent increment was applied to the basic maintenance requirements in the case of light activity under grazing conditions of intensive management and under tropical conditions. A 50 percent increment should satisfy ME requirements on semiarid rangeland pasture and on slightly hilly land. Grazing on sparsely vegetated grassland and on mountainous transhumance pasture, which necessitates daily long-distance travel for grazing and watering and great differences of altitude, may require a 75 percent increment in the basic maintenance needs to meet additional energy requirements. Goats under stable-fed conditions and with minimal activities should be fed according to the basic maintenance requirements only.

Pregnancy

Energy requirements for pregnancy in Table 1 have been derived from the difference of the basic requirements for maintenance (101.38 kcal ME/$W_{kg}^{0.75}$) and two experimental values suggested for pregnancy: 173.60 (Akinsoyinu *et al.*, 1978) and 180.94 (Rajpoot 1979). These values yield a mean of 177.27 kcal ME/$W_{kg}^{0.75}$, which compares closely with 182.41 (McDonald *et al.*, 1973); 179.33 (Akinsoyinu, 1974); and 190.43 (NRC, 1975) for sheep; and 172.97 (NRC, 1978) for dairy cattle. The value of 0.80 Mcal ME in Table 1 also compares well with French recommendations (Morand-Fehr and deSimiane, 1977). In view of the limited information available and the considerable variability among breeds of goats, no differentiation has been made between does producing single kids and those producing twins. An allowance of an additional 20 percent, as recommended for multiple birth in sheep by McDonald *et al.* (1973), has been included. It is understood that pregnancy means the last 2 months of gestation; no extra energy requirements for early pregnancy have been determined. The values take into consideration the weight of the animal and the fetuses.

Growth

Energy requirements for weight gain have been based on three experimental values: 10.18 (Devendra, 1967b); 5.14 (Akinsoyinu, 1974); and 6.43 (Rajpoot, 1979). These values have a mean of 7.25 kcal ME/g of gain, which is equivalent to 4.09 kcal NE. This value is also comparable to those for sheep: 3.74 (Garrett *et al.*, 1959); 3.19 (Evans, 1960); 4.29 (ARC, 1965); 4.21 (Olatungi, 1974); and 2.90 (NRC, 1975). The values for dairy cattle are: 4.19 (Garrett *et al.*, 1959); 2.94 (ARC, 1965); and 3.32 (NRC, 1978). Additional requirements, shown in Table 1, for all growing goats with daily weight gains of 50, 100, and 150 g, have been based on 7.25 kcal ME/g of gain.

Lactation

Energy requirements for lactation have been stated separately for the components of maintenance at different levels of activity and milk production. The requirements have been derived from four experimental values: 1260.00 (Knowles and Watkins, 1938); 1155.90 (Devendra and Burns, 1970); 1328.57 (Winter and Goersch, 1974); and 1240.00 (Rajpoot, 1979). These values yield a mean of 1246.12 kcal ME/kg of 4 percent fat-corrected milk (FCM). This value is comparable to 1202.29 (Blaxter, 1967; McDonald *et al.*, 1973) and 1240.00 kcal ME/kg of 4 percent FCM (NRC, 1978) for dairy cattle. For the requirements of milk production in Table 1 the value of 1246.12 kcal ME/kg of 4 percent FCM has been used over a milk fat range of 2.5 to 6.0 percent. For each 0.5 percent change in fat content from 4 percent milk, an addition or subtraction of 16.28 kcal ME has been applied (NRC, 1978). French recommendations are similar (Morand-Fehr and deSimiane, 1977).

Fiber Production

Energy requirements for fiber production by Angora goats have also been incorporated into Table 1, based on calculations by Huston *et al.* (1971). Although attempts have been made to determine the energy required for mohair fiber production (Gallagher and Shelton, 1972), the values have yet to be established. The values listed in Table 1 were computed by using a 33 percent efficiency of ME for mohair fiber production. No estimates for pashmina production have been made; however, the recommendations for mohair production can be used as a guide.

PROTEIN

Proteins are the principal constituents of the animal body and are continuously needed in the feed for cell repair and synthetic processes. The transformation of feed protein into body protein is an important process of nutrition and metabolism. Proteins consist of amino acids and are the building blocks of all body cells. Secretions such as enzymes, hormones, mucin, and milk have additional amino acid requirements. Proteins are, therefore, vital for animal maintenance, growth, reproduction, and milk production. However, nonprotein nitrogen (NPN) can substitute for parts of the required protein for these functions (Morgen *et al.*, 1907, 1908, 1909, 1910, 1922, 1925; Lawrow *et al.*, 1924; Ungerer, 1924; Paasch, 1925; Honcamp and Keudela, 1927; Williger, 1927; Ziemer, 1931; French, 1957; Champredon and Pion, 1972a,b,c; Fauske, 1972; Kameoka *et al.*, 1972; Mba *et al.*, 1974; Gruhn *et al.*, 1975; Haryu *et al.*, 1975; Harmeyer and Martens, 1980).

Protein deficiencies in the diet deplete stores in the blood, liver, and muscles, and predispose animals to a variety of serious and even fatal ailments. Below a minimum level of 6 percent crude protein (CP) in the diet, feed intake will be reduced, which leads to a combined deficiency of energy and protein (Perkins, 1957; Platt *et al.*, 1964). This deficiency further reduces rumen function and lowers the efficiency of feed utilization. Long-term protein deficiencies retard fetal development, lead to low birth weights, affect kid growth, and depress milk production (Singh and Sengar, 1970).

Protein requirements for maintenance, growth, pregnancy, and lactation have been stated along with the energy requirements in Table 1. They have been presented in terms of total protein (TP) and digestible protein (DP). The former has been recommended as the most accurate guide for converting proteins from feed composition tables to quantities required (Broster, 1972; Preston, 1972; Satter and Roffler, 1975), but DP values are also used widely around the world.

In the past most people have estimated protein requirements based on metabolism trials employing graded protein levels in the diet. The amounts of protein needed to maintain the animal in nitrogen equilibrium were taken as the maintenance requirement. Nitrogen equilibrium was defined as the state in which nitrogen intake matches nitrogen outgo from all sources. The difficulty with this concept is that adult animals can adjust their nitrogen output and reach equilibrium, particularly at lower levels of nitrogen intake. Thus, nitrogen equilibrium as an indicator of adequate protein intake is of questionable value (Singh, 1976) and has not been used here in the determination of protein requirements of goats. Instead, protein requirements have been computed as a ratio to energy.

Some workers have calculated the protein required for maintenance from the endogenous urinary nitrogen (EUN) excretion. Brody (1945) used a factor of 4 to convert EUN to the amount of digestible nitrogen needed for maintenance, while Elliott and Topps (1963) employed a factor of 2.96 for cattle. Neither of these factors can be used satisfactorily for goats; the EUN and maintenance requirements must be determined separately.

Two types of biologically determined protein requirements have been reported. They pertain to minimum and maintenance levels, which must not be confused. The minimum protein requirements for goats have been reported: 1.61 (Majumdar, 1960a); 1.42 (Devendra, 1967a); 1.22 (Singh and Mudgal, 1978); 1.42 (Rajpoot, 1979); and 1.41 (Devendra, 1980b). The mean value is 1.42 g DP or 2.03 g TP/$W_{kg}^{0.75}$ (70 percent average digestibility of total feed protein found in these studies).

Maintenance

Protein requirements for maintenance have been reported: 2.66 (Haenlein, 1950); 2.50 (Majumdar, 1960b); 2.85 (Singh and Sengar, 1970); 3.19 (Winter and Goersch, 1974); 2.12 (Itoh *et al.*, 1979); 3.05 (Rajpoot, 1979); and 3.40 (Sengar, 1980); the mean value is 2.82 g DP or 4.15 g TP/$W_{kg}^{0.75}$, with an average digestibility of 68 percent for total protein. This compares closely with 4.739 TP (NRC, 1975) for sheep and 4.09 g TP/$W_{kg}^{0.75}$ (NRC, 1978) for dairy cattle. For the recommendations in Table 1, the calorie-to-protein ratios were determined to be 1 Mcal DE to 22 g DP to 32 g TP.

Growth

Except for the minimum maintenance requirements of protein, no other biologically determined values for body functions exist. For this report, the experimentally determined energy requirements for maintenance have been used for the development of protein requirements on the basis of calorie-to-protein ratios. In order to complete Table 1, the requirements of protein for weight gains have been derived from the following values: 0.274 (Devendra, 1967b); 0.139 (Akinsoyinu, 1974); and 0.173 (Rajpoot, 1979); the mean is 0.195 g DP or 0.284 g TP/g gain.

Pregnancy

No experimental values have been found for protein requirements of pregnancy. However, two values were calculated from two references: 4.68 (Akinsoyinu *et al.*, 1978) and 4.90 (Rajpoot, 1979); the mean is 4.79 g DP or 6.97 g TP/$W_{kg}^{0.75}$. This value takes into consideration maintenance requirements during the second half of gestation, and is about 10 percent lower than the value of 7.76 g TP/$W_{kg}^{0.75}$ for dairy cattle (NRC, 1978).

Lactation

Protein requirements for milk production were determined from experimental values: 67.83 g DP or 96.90 g TP/kg milk of 4.5 percent fat (Devendra and Burns, 1970) and 46.56 g DP or 66.51 g TP/kg milk of 5.22 percent fat (Rajpoot, 1979), with a mean of 57.20 g DP or 81.71 g TP/kg of milk with 4.86 percent fat. This value is somewhat lower than the 98 g TP/kg of milk with 5 percent fat for dairy cattle (NRC, 1978). In the absence of specific experimental evidence, the protein needs for different milk fat contents have been derived from those recommended for dairy cattle (NRC, 1978). It should be noted that the amounts of protein presented in Table 1 for growth, pregnancy, and lactation are based on calorie-to-protein ratios for maintenance, and therefore cannot be considered accurate for every situation because of differences in reproductive and productive rates. Some flexibility based on experience and judicious judgment is, therefore, advisable. French recommendations (Morand-Fehr and deSimiane, 1977) are similar to those listed in Table 1.

Activity

Protein requirements in Table 1 for muscular activity at the three levels above maintenance have not been determined experimentally. Recommendations for protein (TP and DP) requirements for horses at different stages of work include 26 and 36 percent increments above maintenance for light and medium work for mature animals of 200 to 600 kg body weight (NRC, 1973). This assures that the protein-to-calorie ratio required for maintenance is continued under conditions of work (Crampton, 1964). Later reports indicate that the needs for protein replacement could be small during light work and that excessive protein increases can be undesirable. Supplying greater amounts of concentrates during increased muscular activities may provide additionally needed protein (NRC, 1978). To establish requirements for this report, the TP and DP values for the three activity levels of goats were derived from the protein-to-calorie ratios for the maintenance-only level and low, medium, and high activity levels.

Fiber Production

Protein requirements for fiber production by Angora goats, shown in Table 1, are based on work published at the Texas Agricultural Experiment Station (Huston *et al.*, 1971).

MINERALS

Requirements of minerals have not been established definitively for goats at either maintenance or production levels. However, some classical studies in mineral metabolism have been conducted with goats as experimental subjects. These include studies by Fingerling (1911, 1913), which addressed calcium and phosphorus requirements of lactating goats. Hart *et al.* (1921, 1924, 1927) and Henderson and McGee (1926) reported data on calcium metabolism in goats, which led to the discovery of the role of vitamin D in calcium absorption and metabolism. Lintzel and Radeff (1931) reported important work on iron nutrition in goats. In general, these and more recent studies support assumptions that some mineral requirements in goats are similar to those in other ruminant species; therefore, mineral requirements listed in Table 1 rely for the time being on values recommended for sheep (NRC, 1975) and dairy cattle (NRC, 1978). The literature on mineral nutrition in goats was recently reviewed (Haenlein, 1980).

In addition to the elements in organic matter (oxygen, nitrogen, carbon, and hydrogen), seven major and nine minor minerals are considered dietary essentials for livestock. The major minerals that must be fed in relatively large amounts are calcium, phosphorus, sodium, chlorine, magnesium, potassium, and sulfur. Minor or trace minerals, required in small amounts, include iron, iodine, copper, molybdenum, zinc, manganese, cobalt, selenium, and fluorine. Others which are possibly essential at extremely low levels are chromium, nickel, vanadium, silicon, tin, and arsenic. Most of these essential or possibly essential elements occur naturally in feedstuffs at levels that do not constitute problems in nutrition. However, situations often exist when one or more minerals, especially the major ones, are sufficiently low to reduce productivity. Trace minerals in particular can be present in toxic amounts. Proper balance of minerals and bioavailability from supplements are often more important than actual levels (Miller, 1981). Functions and practical implications of various important minerals are discussed individually.

Calcium

Calcium is a critical nutrient in ration formulation for all species of livestock. Although most of the calcium found in the body is in the skeleton, the element has numerous crucial functions in the soft tissues. A deficiency of calcium in young animals leads to retarded growth and development, and can predispose them to rickets. Because milk is high in calcium (Macy *et al.*, 1953; Parkash and Jenness, 1968), rations for lactating goats need a higher calcium level. Fingerling (1911, 1913) found that if lactating goats did not receive the necessary amounts of calcium and phosphorus in their diets, they would draw from body stores of these elements without initially affecting milk yield or milk composition. If the calcium deficiency continued for weeks, the yield of milk decreased. At intakes of higher levels of calcium the goats replenished their body calcium stores and milk production increased. Changes in milk composition were not observed.

Certain minerals interact with calcium metabolism. Experiments using ligated intestinal loops in anaesthetized goats and radioactively labeled calcium injections (Gibbons *et al.*, 1972) showed that intestinal calcium transport is enhanced by carbohydrates and by low luminal concentrations of sodium. Calcium absorption occurred principally in the duodenum, to a far lesser extent in the jejunum, and least in the lower ileum.

Under grazing conditions calcium is seldom a problem with either Angora or meat-type goats, but it can be very important for high-producing dairy goats. Low calcium diets lead to reduced milk production. Appropriate calcium levels in the diet are also important in the prevention of parturient paresis (milk fever). The calcium content of goats' milk is reported to be in the range of 1.14 to 1.63 g/kg (Macy *et al.*, 1953; Parkash and Jenness, 1968). A median value would be 1.38 g/kg, which is marginally higher than for dairy cattle. This value has been used in formulating the recommendations in Table 1. Suggested calcium supplements or sources include bone meal, dicalcium phosphate, ground limestone, and oyster shell. The percentage composition of these and other sources is found in Table 3.

Phosphorus

Phosphorus is required for both tissue and bone development. A deficiency will result in slowed growth, depraved appetite, and unthrifty appearance; it is often accompanied by low levels of phosphorus in the blood. Fingerling (1911) showed that the general conclusions about calcium-deficient diets also applied to phosphorus. Goats were able to sustain milk production from body reserves for several weeks of negative phosphorus balances. During phosphorus deficiencies, when intake was one-fifth of normal for two months, the production of milk declined 60 percent. Supplementing the diet with P_2O_5 and CaO to achieve daily levels of 6 g phosphorus and 14 g calcium raised milk yields by 10 percent in two weeks and by 15 to 25 percent in four weeks, while diets remained isocaloric and isonitrogenous. The phosphorus level in goats' milk ranges from 0.84 to 1.22 g/kg (Macy *et al.*, 1953; Parkash and Jenness, 1968).

The calcium-to-phosphorus ratio should not drop below 1.2:1 in diets for goats, even though no unanimity exists on the importance of the ratio. The significance of the calcium-to-phosphorus ratio in the genesis of urinary calculi is discussed later on.

A phosphorus deficiency in grazing goats is more likely than a calcium deficiency. It might be encountered with any type of goat grazing on phosphorus-deficient forages. Documented examples of phosphorus deficiency in grazing goats are rare, however. This can be explained by their varied habitats and tendency to browse plants that may be high in phosphorus. Formulation of rations to include adequate phosphorus will be more important with high-producing dairy goats when they are fed with harvested or formulated feedstuffs.

Sodium and Chlorine

Common salt (sodium chloride) is perhaps the mineral most commonly supplied to animals. They require both sodium and chlorine, but sodium is the mineral most likely to be lacking (Schellner, 1972). When provided free choice, goats may consume salt in excess of their requirements, but with no apparent ill effects. Animals that do not receive sufficient salt may show depraved appetites and consume soil or debris. If goats are not provided free choice, salt should be added to the feed. A recommended level would be 0.5 percent of the complete feed or proportionately higher levels in supplements.

Salt is important in several other ways. Placing it in less frequently grazed pastures may influence goats to move to those areas. Salt is also often incorporated at high levels to regulate free intake of nutritional supplements. Trace mineralized salt should not be used in this manner, however; it may lead to an oversupply of some trace elements. Goats in arid regions may have problems with the salt content of some water sources, which can reduce intake of water and feed.

Magnesium

Magnesium is required for many enzyme systems and for proper functioning of the nervous system. It is also closely associated with the metabolism of calcium and phosphorus. Symptoms of magnesium deficiency are anorexia, excitability, and calcification of soft tissue. The most noted problem associated with hypomagnesemia is grass tetany, a malady that frequently occurs in animals grazing on lush green grass or winter cereals in pastures fertilized with nitrogen and potassium. Treatment consists of intravenous administration of calcium and magnesium in the gluconate form. Goats do have a marginal ability to compensate for low dietary magnesium by reducing the rate of its excretion (Razifard, 1971, 1972a,b). Both urinary excretion and milk flow, which contains 0.13 to 0.36 g magnesium per kg, are reduced when magnesium is low in the diet.

Potassium

Potassium, though required in relatively large amounts, is usually present in roughage-based diets to the extent that it does not constitute a problem. Marginal deficiencies result in reduced feed intake, retarded growth, and reduced milk production. More severe deficiencies cause emaciation and poor muscular tone. In growing sheep the potassium requirement is considered to be 0.5 percent of the diet, whereas with lactating dairy cattle the requirement is placed at 0.8 percent of the complete ration (Ward, 1966). These levels are also postulated as the requirements of growing and lactating goats, respectively. Ration values below these are infrequently encountered, and are usually restricted to high-concentrate diets, in which the major ingredient is low in potassium, or to diets of severely weathered or winter range forage. Potassium supplements may be in chloride, bicarbonate, or sulfate form.

Sulfur

Sulfur is a component of all body proteins and is particularly high in goat hair, which consists of a high proportion of the sulfur-containing amino acids, methionine and cystine. Marginal deficiencies cause poor animal performance, and more extreme cases result in excessive salivation, lacrimation, and alopecia. Studies with goats fed supplemental sulfur are rare, but it appears likely that deficiencies of sulfur may be more widespread than previously believed. A recent study by Wheeler *et al.* (1975)

indicates potential shortages of sulfur in forage sorghums. Another study by Gartner and Hurwood (1976) indicates that tannic-acid-containing plants such as *Acacia aneura* may provide inadequate amounts of available sulfur. This is of particular concern with range goats, which liberally graze and browse tannin-containing plants. Recommendations are normally expressed in terms of a sulfur-to-nitrogen ratio of 1:10. However, this ratio may be misleading if either or both sulfur and nitrogen are unavailable because of the presence of complexing substances such as tannic acid. Sulfur requirements would then range from 0.16 to 0.32 percent of the diet for ration protein values of 10 to 20 percent. Sulfates, such as sodium sulfate and ammonium sulfate, are the most available forms of sulfur for ration formulation.

Common feedstuffs may contain adequate sulfur, but shortages can occur in forages grown on certain types of soils or in rations containing a high proportion of NPN as protein supplement (Varma and Sawhusey, 1970). The high-producing Angora goat may have an elevated sulfur requirement because of mohair growth, but this possibility needs to be investigated. It has been shown that sulfur-containing amino acids, administered postruminally, stimulate fiber production (Reis and Schinkel, 1964). Although mechanisms exist, the required technology has not reached the stage of practical application.

Iron

Iron is a component of blood hemoglobin that is required for oxygen transport. It is also required for some enzyme systems. Although iron deficiency seldom occurs in mature grazing animals, it may occur in young goat kids because of their minimal body stores of iron at birth and the low iron content of milk (Jenness, 1980). The work of Lintzel and Radeff (1931) suggests that this may be more true with goats than with cattle. If iron deficiencies are observed and it is desired to continue the kids on a milk diet, injections of iron-dextran (150 mg) at two-to-three-week intervals are recommended. In a recent study (Hamada *et al.*, 1970) acceptable tissue color was observed in animals fed a diet containing 0.03 percent ferrous iron. Thus, this value might be taken as a minimum. Ferrous sulfate and ferric citrate are more available than other sources such as ferric oxide and are recommended for ration formulation. The literature is too sparse, however, to state definite feeding requirements for iron.

Iodine

Iodine is necessary for the formation of thyroxine. In states of iodine deficiency the thyroid gland becomes enlarged, a condition called goiter (Honeker, 1949). It is most frequently observed in the young at birth, especially in weak or dead kids. Iodine-deficient areas are widespread in the world, including parts of the United States. Deficiencies are readily corrected by feeding iodized salt (Sutphin *et al.*, 1971). However, iodized salts should not be force-fed (as when salt is used as a feed limiter) because this action could lead to excessive intakes of iodine.

Goats appear to be somewhat unusual with respect to iodine metabolism (Lengemann, 1970, 1979), but no good basis for unique recommendations exists so far.

Copper and Molybdenum

Copper and molybdenum are interrelated in animal metabolism and should be considered together (Hennig *et al.*, 1974). Levels of both can be too low or too high or the level of one can be low and the other too high. The most common problem occurs when a normal or low level of copper is accompanied by a high level of molybdenum. In this case copper is excreted and a deficiency occurs. This condition can be corrected with copper therapy.

Few studies on copper and molybdenum have included goats (NRC, 1980). It appears that sheep are sensitive to copper toxicity and resistant to molybdenosis, but it is not known whether this is also the case with goats.

Zinc

Zinc deficiency symptoms include parakeratosis, stiffness of joints, excessive salivation, swelling of the feet and horny overgrowth, small testicles, and low libido (Neathery *et al.*, 1973). Reduced feed intake and weight loss also occurs with zinc-deficient diets. Zinc must be supplied continuously because little is stored in the body in readily available form (NRC, 1979). Minimum daily requirements for goats have not been established. Young males have developed deficiencies at levels of 4 ppm (Neathery *et al.*, 1973); adult females developed signs on 6-7 ppm when they were lactating. There is also some evidence that males require more zinc than do females (Groppel and Hennig, 1971; Schellner, 1972). Direct and indirect evidence indicates minimum requirements of 10 ppm. Levels of 1000 ppm may be toxic.

Manganese

Manganese is an essential mineral in diets for goats (Groppel, 1969; Anke *et al.*, 1972, 1973a,b,c; Hennig *et al.*, 1972; Schellner, 1972). Deficiency signs include reluctance to walk, deformity of the forelegs, and reduced reproductive efficiency. So far, data are inadequate to suggest optimum levels. Deficiency signs have developed on 5.5 ppm, but not on 90 ppm in the diet (Anke *et al.*, 1973b).

Other Minerals

Fluorine and selenium can be encountered at either deficient or toxic levels in natural diets. Fluorine deficiency appears to be rare; toxic levels result largely from industrial pollution. With sheep, acute fluorine toxicity occurs at levels above 200 ppm (NRC, 1975). Selenium toxicity occurs in sheep from prolonged consumption of plants containing over 3 ppm. The classical deficiency of selenium is white muscle disease (Hebert and Cowan, 1971), but milder deficiencies result in reduced performance, especially reproductive efficiency. Selenium supplements may be added to salt supplementation or provided through injections.

Cobalt is a component of vitamin B_{12}. Deficiency signs include loss of appetite, emaciation, weakness, anemia, and decreased production. In sheep an intake of 0.1 ppm is considered adequate. It is assumed that the same would apply to goats. Cobalt sulfate or cobalt chloride added at the rate of 12 g per 100 kg of salt should provide an adequate intake, but this would be indicated only in situations in which cobalt has been shown to provide a response.

A few additional references on specific mineral studies with goats can be found in the appended bibliography.

VITAMINS

Vitamins are a group of compounds essential for normal body processes. Typical range or pasture diets of goats should contain adequate levels of vitamins or vitamin precursors to maintain normal health of the animal. Pen-fed animals, goats held on restricted diets, and high-producing animals may need a supplemental vitamin supply (Honeker, 1949). Recommendations in Table 1 for vitamin requirements of goats rely on similar values for sheep (NRC, 1975) and dairy cattle (NRC, 1978) until more specific experimental evidence for goats becomes available.

Vitamin A

Vitamin A is involved in many areas of body metabolism, and as a result deficiency signs are varied. Experimental evidences of vitamin A deficiencies include keratinization of the epithelia of the respiratory, alimentary, reproductive, and urinary tracts, and of the eye. Signs include multiple infections, poor bone development, birth of abnormal offspring, and vision impairment. Night blindness, the inability to see under poorly lighted conditions, is the classic deficiency sign. Experimentally produced signs of a vitamin A deficiency in goats include: loss of appetitie, loss of weight, unthrifty appearance, night blindness, and a thick nasal discharge (Schmidt, 1941).

Vitamin A is not contained in forages, but its precursors are common in plants and are usually present in proportion to plant pigments. However, not all plant pigments give rise to equal vitamin A activity. Beta-carotene is the standard form of provitamin A. One mg of beta-carotene in the diet is equivalent to approximately 400 IU of vitamin A. Other pigments, xanthophylls for example, are less active.

Vitamin A is stored in the liver and fat of animals during times when intake exceeds requirements. During periods of low carotene supplies in the diet, this stored vitamin A can be mobilized and utilized without signs of a vitamin A deficiency. Eveleth *et al.* (1949) reported that a vitamin A deficiency in sheep is unlikely if green feed is available during one season of the year. Schmidt (1941) found the tolerance of adult sheep and Angora goats to low carotene diets to be similar. However, Angora kids were found to be more tolerant than lambs. Goats that have had access to good quality green feed can probably be held on a low carotene diet for a minimum of three months without showing signs of a vitamin A deficiency.

Typical goat diets contain adequate carotene to prevent vitamin A deficiency. The tendency of the goat to search out palatable green plant parts ensures it an advantage over other ruminant species. However, goats that are forced to consume more conventional cattle or sheep diets because of the unavailability of browse would not have an advantage (Davis, 1942; Caldas, 1961). Vitamin A deficiency in goats in the tropics would be rare except under such circumstances.

Old weathered hays are poor sources of carotene, which is readily oxidized. Green leafy hays are good sources, and dehydrated legume hays, especially pelleted, are the best natural sources. Synthetic vitamin A is readily available in feed additive and injectable forms from commercial suppliers. The newer formulations are relatively stable, but old premixes and injectables should not be used.

Vitamin D

Vitamin D is essential for the absorption and metabolism of calcium and phosphorus. In its absence, or at low levels, normal bone development is impaired. Soft, irregular shaped leg and rib bones resulting from a vitamin D deficiency are signs of "rickets." Thus, vitamin D has been referred to as the antirachitic factor. A form of rickets can also be seen in the newborn of an adult female deficient during pregnancy. Otherwise, deficiencies in adult animals are considered rare.

Vitamin D is available to animals both through the diet and as a result of exposure to sunlight. Ultraviolet radiation from sunlight acts on ergosterol, a plant sterol, and on 7-dehydrocholesterol, a sterol of animal origin, to produce compounds having antirachitic activity (vitamin D_2 and D_3, respectively). Thus, sun-cured hays are excellent sources of vitamin D. Animals exposed to sunlight can obtain some of their requirement directly from irradiation of 7-dehydrocholesterol in the skin. DeLuca (1974) discovered that activation of vitamin D_3 occurs in the liver and kidney of animals.

Vitamin D deficiency is unlikely under normal grazing conditions, although a form of osteodystrophia has been produced experimentally in goats. Vitamin D should be supplied to growing animals that are denied sunlight over extended periods because of cloud cover or confinement to housing.

Vitamin E

Vitamin E deficiency in sheep is commonly associated with white muscle disease, also called stiff lamb disease. This malady is seen in young nursing lambs and will improve with vitamin E therapy (Muth *et al.*, 1958). An associated selenium deficiency will intensify the disease. Vitamin E is alleged to improve reproductive efficiency, but dietary supplementation experiments have not produced consistent results.

Evidence of spontaneous vitamin E deficiency signs in goats is lacking. It is suggested that the probability of lowered productivity in goats as a result of a vitamin E deficiency is remote. However, vitamin E transferred to the milk is considered important because of the antioxidant properties that aid in milk storage.

Vitamin K

Vitamin K, the blood clotting vitamin, is plentiful in a variety of feedstuffs and, in addition, is readily synthesized in the rumen. A deficiency is unlikely.

B Vitamin Complex

The B vitamins are not considered dietetically essential in adult goats because they are normally synthesized by microorganisms in the rumen. Only vitamin B_{12} (cobalamin) is likely to be deficient in animals having a functional rumen. Cobalt is required for synthesis of vitamin B_{12} and if absent or at extremely low levels in the diet of goats, a vitamin B_{12} deficiency will occur. The B vitamins should be included in diets of very young kids nursing their dams, animals with poorly functioning rumens, sick animals, and those with radically changed diets.

Vitamin C

Vitamin C is synthesized in the body tissues in adequate quantities to satisfy requirements and under normal circumstances need not be added to diets of goats.

WATER

Water is obviously important for goats, and the amount required depends on that needed for the maintenance of normal water balance and to provide for satisfactory levels of production. The normal body water content of the goat varies with age, amount of fat in the body, and environmental temperatures. It would be expected to exceed 60 percent of the body weight and 75 percent of the nonbony tissues. Shkolnik *et al.* (1980) have shown that some goats, such as the black Bedouin of the Negev and Sinai deserts, have the capacity to store as much as 76 percent of their body weight. Water requirements may be met by free water consumption, but other important sources include water contained in the feed ingested and metabolic water resulting from oxidation of energy sources. Major water losses include those from urine, lactation, evaporation, and perspiration.

A safe general recommendation is to provide goats with all the clean water that they will drink (*ad libitum* intake). Extremes in water temperature will increase energy requirements. Taste factors will also affect normal water intake (Goatcher and Church, 1970). Although the above observations are logical, it should be remembered that a high proportion of the world's goat population lives in areas where water requirements are not easily met. The uniqueness of goats in meeting their water requirements deserves further study. The example of the black Bedouin goat suggests that other genotypes may differ in their ability to meet water requirements. Regardless of breed, water intake must exceed milk production. In a study reported by Bergmann (1932), 3.5 kg of water was consumed for each kilogram of milk produced by dairy goats under temperate conditions. French recommendations are 145.6 g water per $W_{kg}^{0.75}$ for maintenance and 1.43 kg water per kilogram of milk as a production requirement (Morand-Fehr and Sauvant, 1978). In the humid tropics Devendra (1967) found that penned indigenous meat goats had a mean daily free water intake of 680 g, of which 80 percent was consumed during the day.

Goats are often more sensitive and reluctant than other species to drink from foul-tasting water sources. If they are forced to drink poor quality water, the result may be infection or undesirable mineral intake. Also, in many parts of the world goats drink from impounded water, and entrapment (bogging in mud) can be a real hazard, especially with Angoras.

Goats are among the most efficient domestic animals in the use of water, approaching the camel in the low rate of water turnover per unit of body weight (Maloiy and Taylor, 1971; Macfarlane and Howard, 1972). Goats appear to be less subject to high temperature stress than other species of domestic livestock such as wooled sheep or many breeds of cattle and require less water evaporation to control body temperature. They also have the ability to conserve water by reducing losses in urine and feces. In many environments the water intake through forage may be high relative to other species because of their ability or willingness to browse. The result is that goats are less dependent on free water sources than other domestic species, but do not equal certain wild animal species in this respect.

Factors affecting the free water intake of goats are lactation level, environmental temperature, water content of forage consumed, amount of exercise, and salt and mineral content of the diet. Therefore, the daily range of free water intake may be from zero to several liters. When feeding on dry forages and when water is lacking, the efficiency of reproduction will suffer (Brown and Lynch, 1972; Lynch *et al.*, 1972). Suboptimum water intake will result initially in reduced feed intake, then reduced performance and gradual starvation. Acute problems result when goats are unable to maintain water balance or control body temperature.

3 TABLE OF NUTRIENT REQUIREMENTS

The nutrient requirements presented in Table 1 can be used to formulate diets for the different classes and categories of goats by proper use of available feedstuffs. The table lists nutrient requirements for maintenance at different levels of muscular activity and additional requirements for growth, pregnancy, lactation, and mohair production. The daily energy requirements for maintenance are 101.38 kcal $\mathrm{ME}/W_{kg}^{0.75}$ [1 kg TDN = 4.409 Mcal DE (Garrett *et al.*, 1959) or 76 DE = 62 ME = 35 NE; protein requirements are related to energy needs as follows: 22 g DP or 32 g TP per Mcal DE]. The energy requirement for growth is 7.25 kcal ME/g of gain. Goat kids at very young ages may not be able to consume the suggested required quantities of DM. Similar situations can exist with high-producing dairy goats, but intakes of more than 5 percent of body weight have been reported (Haenlein, 1978). The energy requirements per kg FCM at 4 percent are 1246.12 kcal ME; and a 0.5 percent fat change in 4 percent FCM is 16.28 kcal ME. Additional requirements for mohair production are also presented. Details on Table 1 are discussed in Chapter 2.

TABLE 1 Daily Nutrient Requirements of Goats[a]

Body	Feed Energy				Crude Protein				Vitamin A	Vitamin D	Dry Matter per Animal			
											1 kg = 2.0 Mcal ME		1 kg = 2.4 Mcal ME	
Weight (kg)	TDN (g)	DE (Mcal)	ME (Mcal)	NE (Mcal)	TP (g)	DP (g)	Ca (g)	P (g)	(1000 IU)	IU	Total (kg)	% of kg BW	Total (kg)	% of kg BW
Maintenance only (includes stable feeding conditions, minimal activity, and early pregnancy)														
10	159	0.70	0.57	0.32	22	15	1	0.7	0.4	84	0.28	2.8	0.24	2.4
20	267	1.18	0.96	0.54	38	26	1	0.7	0.7	144	0.48	2.4	0.40	2.0
30	362	1.59	1.30	0.73	51	35	2	1.4	0.9	195	0.65	2.2	0.54	1.8
40	448	1.98	1.61	0.91	63	43	2	1.4	1.2	243	0.81	2.0	0.67	1.7
50	530	2.34	1.91	1.08	75	51	3	2.1	1.4	285	0.95	1.9	0.79	1.6
60	608	2.68	2.19	1.23	86	59	3	2.1	1.6	327	1.09	1.8	0.91	1.5
70	682	3.01	2.45	1.38	96	66	4	2.8	1.8	369	1.23	1.8	1.02	1.5
80	754	3.32	2.71	1.53	106	73	4	2.8	2.0	408	1.36	1.7	1.13	1.4
90	824	3.63	2.96	1.67	116	80	4	2.8	2.2	444	1.48	1.6	1.23	1.4
100	891	3.93	3.21	1.81	126	86	5	3.5	2.4	480	1.60	1.6	1.34	1.3
Maintenance plus low activity (= 25% increment, intensive management, tropical range and early pregnancy)														
10	199	0.87	0.71	0.40	27	19	1	0.7	0.5	108	0.36	3.6	0.30	3.0
20	334	1.47	1.20	0.68	46	32	2	1.4	0.9	180	0.60	3.0	0.50	2.5
30	452	1.99	1.62	0.92	62	43	2	1.4	1.2	243	0.81	2.7	0.67	2.2
40	560	2.47	2.02	1.14	77	54	3	2.1	1.5	303	1.01	2.5	0.84	2.1

TABLE 1 *(continued)*

Body Weight (kg)	Feed Energy: TDN (g)	DE (Mcal)	ME (Mcal)	NE (Mcal)	Crude Protein: TP (g)	DP (g)	Ca (g)	P (g)	Vita-min A (1000 IU)	Vita-min D IU	Dry Matter per Animal, 1 kg = 2.0 Mcal ME: Total (kg)	% of kg BW	1 kg = 2.4 Mcal ME: Total (kg)	% of kg BW
50	662	2.92	2.38	1.34	91	63	4	2.8	1.8	357	1.19	2.4	0.99	2.0
60	760	3.35	2.73	1.54	105	73	4	2.8	2.0	408	1.36	2.3	1.14	1.9
70	852	3.76	3.07	1.73	118	82	5	3.5	2.3	462	1.54	2.2	1.28	1.8
80	942	4.16	3.39	1.91	130	90	5	3.5	2.6	510	1.70	2.1	1.41	1.8
90	1030	4.54	3.70	2.09	142	99	6	4.2	2.8	555	1.85	2.1	1.54	1.7
100	1114	4.91	4.01	2.26	153	107	6	4.2	3.0	600	2.00	2.0	1.67	1.7
Maintenance plus medium activity (= 50% increment, semiarid rangeland, slightly hilly pastures, and early pregnancy)														
10	239	1.05	0.86	0.48	33	23	1	0.7	0.6	129	0.43	4.3	0.36	3.6
20	400	1.77	1.44	0.81	55	38	2	1.4	1.1	216	0.72	3.6	0.60	3.0
30	543	2.38	1.95	1.10	74	52	3	2.1	1.5	294	0.98	3.3	0.81	2.7
40	672	2.97	2.42	1.36	93	64	4	2.8	1.8	363	1.21	3.0	1.01	2.5
50	795	3.51	2.86	1.62	110	76	4	2.8	2.1	429	1.43	2.9	1.19	2.4
60	912	4.02	3.28	1.84	126	87	5	3.5	2.5	492	1.64	2.7	1.37	2.3
70	1023	4.52	3.68	2.07	141	98	6	4.2	2.8	552	1.84	2.6	1.53	2.2
80	1131	4.98	4.06	2.30	156	108	6	4.2	3.0	609	2.03	2.5	1.69	2.1
90	1236	5.44	4.44	2.50	170	118	7	4.9	3.3	666	2.22	2.5	1.85	2.0
100	1336	5.90	4.82	2.72	184	128	7	4.9	3.6	723	2.41	2.4	2.01	2.0
Maintenance plus high activity (= 75% increment, arid rangeland, sparse vegetation, mountainous pastures, and early pregnancy)														
10	278	1.22	1.00	0.56	38	26	2	1.4	0.8	150	0.50	5.0	0.42	4.2
20	467	2.06	1.68	0.94	64	45	2	1.4	1.3	252	0.84	4.2	0.70	3.5
30	634	2.78	2.28	1.28	87	60	3	2.1	1.7	342	1.14	3.8	0.95	3.2
40	784	3.46	2.82	1.59	108	75	4	2.8	2.1	423	1.41	3.5	1.18	3.0
50	928	4.10	3.34	1.89	128	89	5	3.5	2.5	501	1.67	3.3	1.39	2.7
60	1064	4.69	3.83	2.15	146	102	6	4.2	2.9	576	1.92	3.2	1.60	2.7
70	1194	5.27	4.29	2.42	165	114	6	4.2	3.2	642	2.14	3.0	1.79	2.6
80	1320	5.81	4.74	2.68	182	126	7	4.9	3.6	711	2.37	3.0	1.98	2.5
90	1442	6.35	5.18	2.92	198	138	8	5.6	3.9	777	2.59	2.9	2.16	2.4
100	1559	6.88	5.62	3.17	215	150	8	5.6	4.2	843	2.81	2.8	2.34	2.3
Additional requirements for late pregnancy (for all goat sizes)														
	397	1.74	1.42	0.80	82	57	2	1.4	1.1	213	0.71		0.59	
Additional requirements for growth—weight gain at 50 g per day (for all goat sizes)														
	100	0.44	0.36	0.20	14	10	1	0.7	0.3	54	0.18		0.15	
Additional requirements for growth—weight gain at 100 g per day (for all goat sizes)														
	200	0.88	0.72	0.40	28	20	1	0.7	0.5	108	0.36		0.30	
Additional requirements for growth—weight gain at 150 g per day (for all goat sizes)														
	300	1.32	1.08	0.60	42	30	2	1.4	0.8	162	0.54		0.45	
Additional requirements for milk production per kg at different fat percentages (including requirements for nursing single, twin or triplet kids at the respective milk production level)														
(% Fat)														
2.5	333	1.47	1.20	0.68	59	42	2	1.4	3.8	760				
3.0	337	1.49	1.21	0.68	64	45	2	1.4	3.8	760				
3.5	342	1.51	1.23	0.69	68	48	2	1.4	3.8	760				
4.0	346	1.53	1.25	0.70	72	51	3	2.1	3.8	760				
4.5	351	1.55	1.26	0.71	77	54	3	2.1	3.8	760				
5.0	356	1.57	1.28	0.72	82	57	3	2.1	3.8	760				

TABLE 1 (*continued*)

Body Weight (kg)	Feed Energy TDN (g)	DE (Mcal)	ME (Mcal)	NE (Mcal)	Crude Protein TP (g)	DP (g)	Ca (g)	P (g)	Vitamin A (1000 IU)	Vitamin D IU	Dry Matter per Animal: 1 kg = 2.0 Mcal ME Total (kg)	% of kg BW	1 kg = 2.4 Mcal ME Total (kg)	% of kg BW
5.5	360	1.59	1.29	0.73	86	60	3	2.1	3.8	760				
6.0	365	1.61	1.31	0.74	90	63	3	2.1	3.8	760				
Additional requirements for mohair production by Angora at different production levels														
Annual Fleece Yield (kg)														
2	16	0.07	0.06	0.03	9	6								
4	34	0.15	0.12	0.07	17	12								
6	50	0.22	0.18	0.10	26	18								
8	66	0.29	0.24	0.14	34	24								

[a] Definitions of terms and equations used are in Chapter 2.

4 HERBAGE AND BROWSE UTILIZATION

In comparison to other domestic animals, goats have unique preferences for shrubs and tree leaves, whether deciduous or evergreen (Fingerling, 1905, 1907, 1909; Cory, 1927; Edwards, 1948; Wilson, 1957). Compared with cattle or sheep they select from a wider array of plants, particularly woody plants (Fraps and Cory, 1940; Maher, 1945; McMahan, 1964). Their pattern of diet selection compares closely with that of small ruminant game animals (McMahan, 1964; Hoppe *et al.*, 1977). Grasses and herbaceous flowering plants (forbs or weeds) are also commonly selected (Knight, 1965; Malechek and Leinweber, 1972a,b; Tetteh, 1974). Wilson (1957) observed that goats showed a special preference for the influorescences of grasses. Because of their unusual preferences for leaves of woody plants, they have been exploited as weapons against encroaching brush species. Early studies in the deciduous woodlands of the United States observed that goats would effectively clear away understory and would actually kill some trees up to 6 inches in diameter (Woods, 1903). In East African thornbush areas, goats controlled sprouts and regrowth following mechanical control (Staples *et al.*, 1942; Oates, 1956). However, goats do not select only invading plants, and care must be taken to avoid the overstocking that can lead to destruction of all ground cover (Hornby and Rensburg, 1948; Wilson *et al.*, 1975). The most successful use of goats in the control of invading plants involves intensive grazing for a short period, followed by removal of all grazing for an extended period to allow for recovery of desirable plants. This method is not, of course, effective when goats find invading plants unpalatable.

Provenza (1978) portrayed the goat as a mobile pruning machine that modified bushy shrubs and thereby increased the accessibility of cattle to more nutritious forage. The observation of synergistic effects between animal species has led to widespread acceptance of combination grazing. Talbot and Talbot (1963) recognized the mutual benefit of mixed species grazing by wild ruminants in East African savannas. Mixed grazing has also increased yields of animal products from rangelands in Texas (Merrill, 1975).

Goats consume approximately the same weight of forage DM as do sheep of similar size (Geoffroy, 1974). The exact amount that they will voluntarily consume is influenced by several factors. Malechek and Leinweber (1972a,b) suggest that goats will eat more forage if they have access to the more preferred species. Devendra (1975) found that voluntary intake by goats decreased as the forage matured. This effect is overcome partially by chopping and pelleting the forage (Fehr, 1971; Fonolla *et al.*, 1972; Devendra, 1977b). Environmental factors such as temperature and humidity also often affect the level of voluntary intake (Chenost, 1972b).

Browse (leaves and twigs of trees and shrubs) and forbs generally contain higher levels of crude protein and phosphorus during the growing season than do grasses (Rector and Huston, 1976). But many palatable browse species are limited in value because of one or more inhibitors that may bind or otherwise prevent utilization of nutrients contained in the plants. These inhibitors include excessive lignification of woody twigs, and tree leaves that physically bind or encapsulate the nutrients (Short and Reagor, 1970; Singh *et al.*, 1972). Essential oils (terpene-based organic compounds) are present at relatively high levels in some range shrubs and apparently inhibit growth of rumen bacteria (Oh *et al.*, 1968). High levels of tannins are found in some of the important browse plants and depress digestion of feedstuffs by binding and/or inhibiting enzymatic activity (McLeod, 1974; Gartner and Hurwood, 1976). Fraps (1924) reports the trace of digestible protein content of live oak leaves was zero. Nastis (1977) found Gambel oak (*Quercus gambelii*) to be less digestible at a young, tender stage than when mature because of the high tannin concentration in young leaves. Other compounds such as silica may also limit the value of browse materials (Short *et al.*, 1973). Although the reasons are not fully understood, field observations indicate that goats under browse conditions perform better than might be expected from the above cited reports. In spite of concern that browse materials are less nutritious than would be indicated by chemical analysis, there is growing evidence that many, if not most, of the grasses, shrubs, and tree leaves selected by goats are of high nutritional value (Fraps and Cory, 1940; Bissell and Weir, 1957; Butterworth, 1967; Wilson *et al.*, 1971; Bhandari and Gupta, 1973; Short *et al.*, 1974; Cordova and Wallace, 1975; Rector and Huston, 1976; Wilson, 1977). It has been suggested that browse plants, shrubs, and tree leaves should be given more attention in the feeding management of goats (Sidahmed *et al.*, 1981b).

Several reports indicate that goats are more efficient digesters of forage, especially the fiber fraction, than other domestic animals (Ademosun, 1970a,b; Gihad, 1976; Devendra, 1977a,b; Sharma and Rajora, 1977). Huston (1976) reported a lower digestive efficiency for goats than for cattle for several forages. Other workers have shown that goats surpass other ruminants in digesting some forages but not others (Jones *et al.*, 1972; ElHag, 1976). Nonplant factors affecting digestibility of forages by goats include geographical location (Chenost, 1972a), level of forage consumption (Devendra, 1967a,b; Sharma and Murdia, 1974), concentrate supplementation (Castle, 1956a; Chenost, 1972c; Saxena *et al.*, 1972), and rate of passage through the gastrointestinal tract (Castle, 1956b; Short *et al.*, 1965; Ehrlein and Hill, 1970; Hamada, 1973). Because significant differences are usually small, it is suggested at this stage that the digestive efficiencies of sheep and goats be considered comparable. Any differences that appear to exist deserve further studies (Coblentz, 1977; McCammon-Feldman, 1980; Van Soest, 1980; Sidahmed *et al.*, 1981a,b).

Campbell *et al.* (1962) conceded the importance of the brush-eating characteristics of the goat, but emphasized the animal's importance as a utilizer of forages and a producer of human consumable goods (meat, milk, and fiber). The goat offers an opportunity, sometimes the only alternative, for deriving value from a vast reservoir of natural resources, and unwanted assortments of herbage, shrubs, tree leaves, and plant refuse and by-products.

5 RATION FORMULATION AND EXAMPLES OF TYPICAL RATIONS

Formulating practical rations requires (1) identification of the total nutrient requirements of a goat or herd of goats according to body size and the physiological functions to be satisfied, and (2) proper combination of available feedstuffs to supply these nutrients in the most economical manner. The requirements listed in Table 1 can be met in many different ways using different combinations of feed ingredients. Local availability and cost of ingredients will be the determining factors. Roughages are normally the least expensive and are considered first. The remaining requirements are met with concentrates. Requirements of goats are presented in Table 1 for:

1) Maintenance
2) Growth and live weight gain
3) Pregnancy
4) Milk production
5) Mohair production

Included with maintenance is a consideration of activity levels that are considered as part of "practical maintenance." These increased levels result from foraging activity, travel, and climbing. Perhaps more than any domestic animal species, goats vary in degree of activity with breed, location, management system, and climate. Thus, four different levels of requirements for "practical maintenance" are considered:

1) Maintenance only, including minimal activity
2) Maintenance plus low level activity
3) Maintenance plus medium level activity
4) Maintenance plus high level activity

The maintenance requirement used to calculate ration examples assumes a largely minimal level of activity, but appropriate rations for goats in more active circumstances can be computed from the respective data for other levels in Table 1. Feeds in the temperate regions will vary considerably from those in the tropics, but a few typical feeds have been selected. While alfalfa hay, corn, and soybean oilmeal may be fed frequently in the temperate regions, in the tropics one might use rather Guinea grass, Cassava chips, coconut cake, and fish meal. Roughages include browse, tree leaves, and crop residues; their voluntary intake will be affected by the dry matter contents. Nutrient contents for all ration ingredients can be extracted from Tables 2 and 3. Rations satisfying all requirements can be calculated not only for energy and protein demands, as in the examples below, but also for minerals and vitamins, as far as nutrient needs are known on a minimum or a challenge basis of supply. Chapter 2 gives additional details for the use of Table 1. Ration formulation examples are based on Tables 2 and 3, but in many instances the data are inadequate or needs have not yet been established specifically for goats. Fortunately, legume forages are good sources of most mineral elements required by goats. When high-quality legumes make up half or more of the roughage in a dairy goat ration, as may be often the case in temperate countries, most of the mineral requirements, except possibly selenium, may be met. The situation can be different in tropical countries where fodder grass, with or without tree leaves, rather than legumes is more commonly fed, and additional mineral needs have to be satisfied. With dependence on legumes, however, phosphorus and sodium may be deficient enough that supplementation will be necessary. The sodium requirement can usually be fulfilled by including 0.5 percent salt (sodium chloride) in the concentrate mixture, or by providing it *ad libitum* as granular salt or salt block.

When poor quality roughages including legumes, forages, and tree leaves are predominant in a ration, more extensive mineral supplementation with calcium, phosphorus, manganese, zinc, copper, iron, selenium, iodine, and cobalt is required. Where possible, all forages should be tested for their mineral content because extreme variation is common, particularly with tropical feedstuffs.

Users of the following ration examples should be aware that the suggested intake levels may not always be achieved in young kids and high-producing dairy goats, although intakes of more than 5 percent body weight have been reported (Haenlein, 1978b). Furthermore, it must be recognized that goats prefer selecting among and within various feed sources and that more forage is often needed in order to achieve the indicated net intake levels.

1. EXAMPLE RATIONS FOR MAINTENANCE

A. *For a 30 kg goat in tropical areas in a nonproductive state with maintenance only and minimal activity.*

Total Requirements (from Table 1): 1.59 Mcal DE/day
51.0 g TP/day

Ration

	DM basis			As-fed basis		
Feeds	Amount (g)	DE (Mcal)	TP (g)	DM (%)	Amount (g)	% of ration
Chickpea (Gram), straw	620	1.36	33	91	681	65
Alfalfa, fresh	95	0.23	19	26	365	35
Total	715	1.59	52	—	1,046	100

Composition of ration: DE = 2.22 Mcal/kg DM
TP = 7.3% of DM

Level of intake (DM): 2.4% of body weight

B. *For a 50 kg goat in tropical areas in a nonproductive state with maintenance only and minimal activity.*

Total Requirements (from Table 1): 2.34 Mcal DE/day
75 g TP/day

Ration

	DM basis			As-fed basis		
Feeds	Amount (g)	DE (Mcal)	TP (g)	DM (%)	Amount (g)	% of ration
Wheat straw	716	1.40	26	89	804	30
Alexandrian clover (Berseem), fresh	333	0.94	56	18	1,850	70
Total	1,049	2.34	82	—	2,654	100

Composition of ration: DE = 2.23 Mcal/kg DM
TP = 7.8% of DM

Level of intake (DM): 2.1% of body weight

2. EXAMPLE RATIONS FOR LIVEWEIGHT GAIN

A. *For a 20 kg growing animal with minimal body activity gaining 50 g per day.*

Total Requirements (from Table 1)

Maintenance	1.18 Mcal DE/day	38 g TP/day
Growth	0.44	14
Total	1.62	52

Ration

	DM basis			As-fed basis		
Feeds	Amount (g)	DE (Mcal)	TP (g)	DM (%)	Amount (g)	% of ration
Alfalfa hay, full bloom	80	0.19	14	19	88	18
Corn grain	360	1.44	38	87	414	82
Total	445	1.66	52	—	502	100

Composition of ration: DE = 3.73 Mcal/kg DM
TP = 11.7% of DM

Level of intake (DM): 2.2% of body weight

B. *For a 30 kg growing goat with minimal body activity gaining 150 g per day.*

Total Requirements (from Table 1)

Maintenance	1.59 Mcal DE/day	51 g TP/day
Growth	1.32	42
Total	2.91	93

Ration

	DM basis			As-fed basis		
Feeds	Amount (g)	DE (Mcal)	TP (g)	DM (%)	Amount (g)	% of ration
Chickpea straw	500	1.10	26	91	549	51
Corn grain	400	1.60	42	87	460	42
Linseed oilmeal	65	0.23	25	90	72	7
Total	965	2.93	93	—	1,081	100

Composition of ration: DE = 3.03 Mcal/kg DM
TP = 9.6% of DM

Level of intake (DM): 3.2% of body weight

3. EXAMPLE RATIONS FOR PREGNANT DOES

A. *For a 30 kg doe in late gestation and having minimal activity.*

Total Requirements (from Table 1)

Maintenance	1.59 Mcal DE/day	51 g TP/day
Pregnancy	1.74	56
Total	3.33	107

Ration

	DM basis			As-fed basis		
Feeds	Amount (g)	DE (Mcal)	TP (g)	DM (%)	Amount (g)	% of ration
Wheat straw	500	0.98	18	89	562	25
Oat silage	365	1.00	35	30	1,216	55
Barley grain	400	1.44	53	90	444	20
Total	1,265	3.42	106	—	2,222	100

Composition of ration: DE = 2.70 Mcal/kg DM
TP = 8.4% of DM

Level of intake (DM): 4.2% of body weight

B. *For a 40 kg doe in late gestation and with minimal activity.*

Total Requirements (from Table 1)

Maintenance	1.98 Mcal DE/day	63 g TP/day
Pregnancy	1.74	56
Total	3.72	119

Ration

	DM basis			As-fed basis		
Feeds	Amount (g)	DE (Mcal)	TP (g)	DM (%)	Amount (g)	% of ration
Johnsongrass hay, mature	960	2.36	91	91	1,055	73
Sorghum grain	350	1.35	40	89	393	27
Total	1,310	3.71	131	—	1,448	100

Composition of ration: DE = 2.83 Mcal/kg DM
TP = 10.0% of DM

Level of intake (DM): 3.3% of body weight

4. EXAMPLE RATIONS FOR LACTATING DOES

A. *For a 30 kg doe producing 1 kg of milk testing 4% fat and having minimal activity.*

Total Requirements (from Table 1)

Maintenance	1.59 Mcal DE/day	51 g TP/day
Lactation	1.53	72
Total	3.12	123

Ration

	DM basis			As-fed basis		
Feeds	Amount (g)	DE (Mcal)	TP (g)	DM (%)	Amount (g)	% of ration
Alexandrian clover (Berseem), hay	500	1.42	76	88	568	49
Molasses, cane	200	0.70	12	74	270	23
Cassava, chips	200	0.79	7	81	247	21
Peanut oilmeal	60	0.21	31	92	65	7
Total	960	3.12	126	—	1,150	100

Composition of ration: DE = 3.25 Mcal/kg DM
TP = 13.1% of DM

Level of intake (DM): 3.2% of body weight

B. *For a 70 kg goat producing 5 kg of milk testing 3.5% fat and with minimal activity.*

Total Requirements (from Table 1)

Maintenance	3.01 Mcal DE/day	96 g TP/day
Lactation	7.55	340
Total	10.56	436

Ration

	DM basis			As-fed basis		
Feeds	Amount (g)	DE (Mcal)	TP (g)	DM (%)	Amount (g)	% of ration
Corn silage, dough stage	1,000	2.92	77	27	3,704	60
Alfalfa hay, full bloom	500	1.18	85	91	549	9
Corn grain	1,365	5.45	145	87	1,569	26
Soybean oilmeal	280	1.09	130	90	311	5
Total	3,145	10.64	437	—	6,133	100

Composition of ration: DE = 3.38 Mcal/kg DM
TP = 13.9% of DM

Level of intake (DM): 4.5% of body weight

C. *For a 60 kg goat producing 6 kg of milk testing 3.5% fat and having a low level of activity.*

Total Requirements (from Table 1)

Maintenance	3.35 Mcal DE/day	105 g TP/day
Lactation	9.06	408
Total	12.41	513

Ration

	DM basis			As-fed basis		
Feeds	Amount (g)	DE (Mcal)	TP (g)	DM (%)	Amount (g)	% of ration
Mixed grass hay	500	1.30	38	89	562	14
Corn grain	1,700	6.78	180	87	1,954	50
25% protein supplement	1,200	4.40	300	85	1,412	36
Total	3,400	12.48	518	—	3,928	100

Composition of ration: DE = 3.67 Mcal/kg DM
TP = 15.2% of DM
Level of intake (DM): 5.7% of body weight

5. EXAMPLE RATIONS FOR ANGORA GOATS

A. *For a 30 kg nonpregnant, nonlactating doe having medium activity and producing mohair at a rate of 4 kg per year.*

Total Requirements (from Table 1)

Maintenance	2.38 Mcal DE/day	74 g TP/day
Mohair	0.15	17
Total	2.53	91

Ration

	DM basis			As-fed basis		
Feeds	Amount (g)	DE (Mcal)	TP (g)	DM (%)	Amount (g)	% of ration
Johnsongrass hay	600	1.44	46	91	659	67
Sorghum grain	252	0.97	29	89	283	29
Cottonseed oilmeal	34	0.13	16	90	38	4
Total	886	2.54	91	—	980	100

Composition of ration: DE = 2.87 Mcal/kg DM
TP = 10.3% of DM
Level of intake (DM): 3.0% of body weight

B. *For a 20 kg goat kid gaining 100 g/day, having low body activity and producing mohair at a rate of 2 kg per year.*

Total Requirements (from Table 1)

Maintenance	1.47 Mcal DE/day	46 g TP/day
Growth	0.88	28
Mohair	0.07	9
Total	2.42	83

Ration

	DM basis			As-fed basis		
Feeds	Amount (g)	DE (Mcal)	TP (g)	DM (%)	Amount (g)	% of ration
Alfalfa hay, mature	250	0.59	33	91	275	32
Corn grain	441	1.76	47	87	507	60
Molasses, cane	50	0.17	3	74	68	8
Total	741	2.52	83	—	850	100

Composition of ration: DE = 3.40 Mcal/kg DM
TP = 11.2% of DM
Level of intake (DM): 3.7% of body weight

C. *For a 40 kg pregnant doe having low body activity and producing mohair at a rate of 6 kg per year.*

Total Requirements (from Table 1)

Maintenance	2.47 Mcal DE/day	77 g TP/day
Pregnancy	1.74	56
Mohair	0.22	26
Total	4.43	159

Ration

	DM basis			As-fed basis		
Feeds	Amount (g)	DE (Mcal)	TP (g)	DM (%)	Amount (g)	% of ration
Alfalfa hay, mature	700	1.66	93	91	769	48
Corn grain	630	2.51	67	87	724	46
Molasses, cane	75	0.26	4	74	101	6
Total	1,405	4.43	164	—	1,594	100

Composition of ration: DE = 3.15 Mcal/kg DM
TP = 11.7% of DM
Level of intake (DM): 3.5% of body weight

D. *For a 30 kg doe having high body activity, nursing at the rate of 1 kg of milk production of 4% fat per day, and producing mohair at a rate of 4 kg per year.*

Total Requirements (from Table 1)

Maintenance	2.78 Mcal DE/day	74 g TP/day
Lactation	1.53	72
Mohair	0.07	9
Total	4.38	155

Ration

	DM basis			As-fed basis		
Feeds	Amount (g)	DE (Mcal)	TP (g)	DM (%)	Amount (g)	% of ration
Johnsongrass hay	400	0.98	30	91	440	28
Alfalfa hay, mature	400	0.95	53	91	440	28
Corn grain	600	2.39	64	87	690	42
Cottonseed oilmeal	20	0.08	9	91	22	2
Total	1,420	4.40	156	—	1,592	100

Composition of ration: DE = 3.1 Mcal/kg DM
TP = 11.0% of DM
Level of intake (DM): 4.7% of body weight

6. OTHER EXAMPLE RATIONS FOR GOATS IN TEMPERATE AND TROPICAL REGIONS AND FOR ANGORA GOATS

A. *Temperate regions*

Does: pregnant or dry
Example 1:
pasture plus good mixed hay and
0.5 kg of a 16% protein supplement
Example 2:
0.5 kg silage
0.5 kg mixed hay
0.3 kg beet pulp
0.5 kg 16% protein supplement
Example 3:
1.0 kg beets
0.5 kg alfalfa hay
0.5 kg beet pulp
0.5 kg 16% protein supplement

Does: lactating
Example 1:
1.5 kg clover hay
2.0 kg 14% protein supplement
Example 2:
1.5 kg grass legume hay
2.5 kg 16% protein supplement
Example 3:
0.5 kg mixed hay
2.5 kg corn silage
2.0 kg 18% protein supplement
Example 4:
3.0 kg roots, beets, carrots, steamed potatoes
1.5 kg mixed hay
0.25 kg beet pulp
0.5 kg oats straw
1.0 kg 14% protein supplement
Example 5:
2.0-4.0 kg green chop, pasture
1.5 kg sugar beet leaf silage
0.5 kg alfalfa hay
0.7 kg beet pulp
0.45 kg 14% protein supplement

Kids: nursing
Colostrum on the 1st day, 0.25 to 1.0 kg milk 2 to 3 times a day according to size for six to nine weeks, plus 16% protein supplement consisting of coarse grain, steamed rolled corn, oats, barley, pelleted alfalfa leaf meal, molasses (not more than 10%), and grass hays *ad libitum*

Kids: weaned and yearlings
Good mixed hay *ad libitum*, plus 0.25 to 0.75 kg of 16% protein supplement consisting of coarse grain mixtures and pasture

Bucks: breeding
(out of season)
Good hays *ad libitum* and pasture
(in season)
0.5 to 1.0 kg of a 14% protein supplement, plus mineral supplementation and salt, plus good hays and pasture

B. *Tropical regions*

Postweaning growth and meat production:
Example 1 (India):
50% cereal straw
30% corn grain
20% Alexandrian clover (Berseem), green

Example 2 (India):
- 40% cereal straw
- 30% oat silage
- 25.5% corn grain
- 4.5% linseed oilmeal

Example 3 (Nigeria):
- 79.3% cassava flour
- 15.2% molasses
- 5.5% urea

Pregnant doe (India):
- 40% cereal straw
- 30% oat silage
- 23.5% barley grain
- 6.5% peanut oilmeal

Supplements for milk production (fed at pasture):

Example 1 (India):
- 45% corn grain
- 35% Alexandrian clover, (Berseem)
- 20% cereal straw

Example 2 (West Indies):
- 34% coconut meal cake
- 20% wheat middlings
- 20% molasses
- 15% citrus meal
- 10% soybean oilmeal
- 1% mineral mix

Example 3 (Malaysia):
- 40% wheat flour
- 34% rice bran
- 12% peanut oilmeal
- 10% coconut meal cake
- 2% molasses
- 2% mineral mix

Example 4 (Mexico):
- 44% sorghum grain
- 37% corn grain
- 10% soybean oilmeal
- 6% molasses
- 1% urea
- 2% salt and mineral mix

C. *Angora goats*

Growing kids and yearlings:
- 32% alfalfa hay
- 28% cottonseed hulls
- 18% sorghum grain
- 8% barley grain
- 6% molasses
- 6% cottonseed oilmeal
- 2% salt and mineral mix

Lactating does:
- 47% alfalfa hay
- 20% cottonseed hulls
- 15% sorghum grain
- 8% barley grain
- 6% molasses
- 2% cottonseed oilmeal
- 2% salt and mineral mix

7. EXAMPLE PROTEIN SUPPLEMENTS (PERCENT OF RATION)

	Total protein content		
	14%	16%	18%
Corn grain	37	35	32
Oats grain	37	35	32
Wheat bran	16	14	15
Oilmeal, soybean, linseed	9	15	20
Dicalcium phosphate	0.5	0.5	0.5
Trace mineral salt	0.5	0.5	0.5
Total	100	100	100

6 NUTRITION-RELATED METABOLIC DISORDERS

ABORTION

The goat is more susceptible to abortion than other species of domestic livestock. Most of the work relating to abortion in goats has been with the Angora (Van Heerden, 1963; Van Rensburg, 1971; Shelton and Groff, 1974), in which the problem is more severe. Infectious diseases such as brucellosis are also capable of causing abortion in goats (Alton, 1973). The fact that the goat is a *corpus-luteum*-dependent species predisposes the animal to abort whenever there is an interference with a functional *corpus luteum* (Wentzel *et al.*, 1975). A low level of abortion is common with the Angora under normal production conditions, but catastrophic losses sometimes occur. Most abortions occur in response to stress between 90 and 110 days of gestation. Undernutrition during the critical stage of rapid fetal development and competition for nutrients between fetal and maternal organisms appear to be one explanation. The incidence of abortion is reduced in flocks in which replacement does are fed for proper size and development prior to the first breeding season and during gestation (Shelton and Stewart, 1973).

A series of studies from South Africa appears to provide a physiological explanation for the type of abortion observed in that country with the Angora. Parturition, either at or prior to term, is normally initiated by elevated corticosteroids of fetal or maternal origin (Wentzel and Roelofse, 1975). Two types of abortion have been identified in the Angora. One is known as stress abortion, which is triggered by low maternal blood glucose (Wentzel *et al.*, 1976). This type is normally induced by poor nutritional condition of the doe (Wentzel *et al.*, 1974), but other stress factors are also involved. Stress abortion is identified by the expulsion of a live or fresh fetus. Low maternal glucose appears to trigger hyperactivity of the fetal adrenal. The cause of abortion in the period 90-110 days of pregnancy is apparently explained by the fetal adrenal gland's producing elevated levels of estrogen precursors (Wentzel *et al.*, 1976), and estrogens are known to be potent abortifacients (Wentzel *et al.*, 1975). After 110 days the fetal adrenal is more mature and produces corticosteroids, which are slower acting or less potent abortifacients. A second type of abortion is that by the habitual aborter. These goats can be identified by a history of abortion, and by the expulsion of a dead edematous or autolyzed fetus. This type of abortion apparently results from maternal hyperadrenalism. Both types of abortion may be triggered by undernutrition resulting in low blood glucose. Initial or stress abortions can be almost totally prevented by adequate nutrition and the elimination of stress.

ENTEROTOXEMIA

It has been said with reason that it is impossible to manage a herd of good dairy goats without experiencing some incidence of enterotoxemia, also known as toxic indigestion or overeating disease (Guss, 1977). Diarrhea, depression, lack of coordination, digestive upsets, coma, and death may be observed after excessive feeding on the part of both baby kids and mature animals. Excessive feeding may occur after sudden changes in feeds; with access to palatable, readily fermentable feeds relished by hungry goats; and under conditions of calcium insufficiency and acidosis. Enterotoxemia is a toxic reaction to *Clostridium perfringens* type C or D, against which antitoxins and vaccination programs with toxoid or bacterins are effective. However, the best prevention in stable-fed goats is frequent feeding of milk, grain, and forage in small amounts. Large meals given once a day should be avoided. Changes of concentrates and forages in the ration should be introduced gradually over several days, especially when the protein or energy content of the diet is increased. When urea or other nonprotein nitrogen is to be part of the diet, then the gradual adaptation should take at least three weeks.

Acute indigestion with a rumen pH of less than 4.8 indicates lactic acidosis. It can follow high levels of grain feeding in early lactation and may lead to the secondary complication of enterotoxemia. Recent research with sheep and cattle on the sensitivity of *Streptococcus bovis*, the initiator of acute acidosis, to various antibiotics gives hope that powerful aids in the prevention of enterotoxemia may be available for goats (Muir *et al.*, 1981).

KETOSIS

Ketosis is a metabolic disorder defined by increased levels of ketone bodies (acetone, betahydroxybutyric acid, and acetoacetic acid) in blood, milk, and urine, and is associated with elevated blood plasma nonesterified fatty acids (NEFA), which are precursors of ketone bodies. Lactation ketosis is observed primarily in high-producing dairy cows and to a lesser extent in dairy goats (Leach, 1971; Mackenzie, 1973; Schultz, 1974; Guss, 1977). Late-pregnancy ketosis is encountered in sheep and goats carrying multiple fetuses. Goats appear to be more resistant than cows or ewes to ketosis. Treatment is similar to that for cows: intravenous glucose, glucocorticoid steroids, adrenocorticotrophic hormone (ACTH) injections, oral drenching with sodium propionate, propylene glycol or chloral hydrate.

Experimentally, phlorizin injections in goats have simulated ketosis by causing glucosuria, hypoglycemia, ketonemia, and blood plasma NEFA level increases (Menahan, 1966). Forcing goats to go without food also results in increases of blood NEFA levels, especially in late pregnancy or during lactation. However, the fat depot is the ultimate source of ketosis, and dairy goats appear to lack the fat reserves of cows and other animals, which might explain why ketosis is unusual in goats.

Increased ketogenesis has been produced by infusion of butyric acid into the rumen of goats with phlorizin-induced hypoglycemia; the condition was corrected by intravenous injections of glucose or propionate, or intraruminal administration of propionate (Menahan, 1966). Most ketogenesis was produced, however, from butyric acid infusion into the rumen in late pregnancy and when the goats were forced to fast.

Diabetes accompanied by elevated blood NEFA levels has also been simulated in goats by the administration of intravenous alloxan (Menahan, 1966). A glucose drain during late multiple pregnancy and heavy lactation is the triggering stimulus to lipolysis and ketogenesis. A feedback effect from ketonemia in the presence of insulin appears to prevent further increases of fat mobilization and may be important to the survival of the animal.

Increased plasma NEFA levels were a more sensitive indicator of undernutrition in goats than blood ketones or blood glucose levels (Radloff, 1964). Growth hormone, epinephrine, glucocorticoid steroids, and ACTH have direct effects on and relationships to blood ketone levels in goats.

PARTURIENT PARESIS (MILK FEVER)

The incidence of paralysis-type conditions of this metabolic hypocalcemic disorder differs among genetic groups of dairy cattle, but is also reported for other species, including goats (Littledike, 1974; Guss, 1977). However, it is not observed as frequently in goats as in cows. Signs and treatments are similar to those for cows. Prevention has been tried with different contents of calcium in the diet during the dry period, and with hormone treatment and vitamin D therapy, but no generally accepted management practice has evolved. Parturient paresis has been related to greatly increased mammary blood flow immediately after parturition (Reynolds, 1970).

The relationship of the kinetics of calcium pool size and calcium turnover rate to dietary phosphorus levels were studied and discussed by Twardock *et al.* (1970) and Anderson *et al.* (1970). It was noted that different dietary regimes of goats, including changed calcium-to-phosphorus ratios, had significant effects on the size and biological half-life of the readily exchangeable calcium pool. It was suggested that the response time of the parathyroid hormone and the removal of calcium from the so-called nonexchangeable bone pool was too slow for the immediate calcium needs of parturition and lactation onset in goats; and when the readily exchangeable calcium pool was inadequate to meet these needs, then parturient paresis resulted. A low or high calcium diet over time may be a predisposing factor for a reduced readily exchangeable calcium pool in goats. Dietary phosphorus levels influence the effects of such diets, and also the level of intestinal calcium absorption and available calcium in goats.

POSTHITIS

Posthitis, also known as sheathrot or pizzlerot, has been reported in male goats (Shelton and Livingston, 1975). This problem has been studied extensively in Merino wethers in Australia (Osborne and Widdows, 1961; McMillan and Southcott, 1973). The causative agents are thought to be a high protein ration in combination with the presence of a urea-hydrolyzing organism such as *Corynebacterium renale* (McMillan and Southcott, 1973; Barajas and Biberstein, 1974; Shelton and Livingston, 1975). The problem appears to be aggravated by confinement to areas where irritation or infection are more likely to occur. The problem is not likely to be a serious one with goats, except with mature Angora wethers kept for hair production. The problem may also occur with individual breeding bucks kept in confinement.

TOXIC PLANTS

The problem of poisonous plants is of great importance to owners of much of the world's goat population. It is not known, nor can it be inferred, that goats are either more or less susceptible to toxic plants than other animal species. However, their grazing habits and the environment under which many of them are kept place them in wider contact with toxic plants. Many goats are found in arid areas, and are noted for eating a variety of plants under these conditions. Many native forage species found under arid conditions have natural protective mechanisms, including toxic principals that retard evaporation and protect against livestock. Goats are thought to be less sensitive than cattle to the toxic effects of tannic acid. Goats can live for extended periods of time on oak species with high tannic acid, whereas cattle are very susceptible to this material (Dollahite, 1961). Goats are also not bothered by bitterweed (*Hymenoxys odorata*), which causes severe losses with sheep in some areas (Hardy *et al.* 1931). And goats have been used at times to reduce the availability of toxic plants to other animal species (Dollahite, 1972). Some references to toxic plants and their effects include Sperry *et al.* (1964), Kingsbury (1964), Lindahl (1972), and Keeler *et al.* (1978).

UREA TOXICITY

Urea is an important natural compound in the physiological processes of goats, but can be highly toxic if consumed in excess. Although most of the urea that is formed in the liver is excreted through the kidney, a portion passes into the rumen where it is hydrolyzed to ammonia and used by rumen microorganisms for protein synthesis (Vercoe, 1969; Hume *et al.*, 1970). Therefore, urea is frequently included in ruminant diets to partially replace protein ingredients. Producers and feed formulators must exercise caution when feeding goats urea, since excessive amounts can result in a buildup of ammonia to toxic levels in the bloodstream (Morris and Payne, 1970; Kromann *et al.*, 1971). It is recommended that urea supply no more than one-third of the total crude protein in forage or roughage-type diets and not more than one-half in the concentrate portion of the diet. Also, an adaptation period of at least three weeks is required for the animal to utilize urea efficiently. It is generally believed that 44 g/100 kg body weight at a single feeding will result in acute toxicity. Producers should assure that daily consumption levels at that rate do not occur.

UROLITHIASIS

Goats are known to be susceptible to urolithiasis (urinary calculi), and serious losses can occur when valuable breeding males are placed on calculogenic rations (Sato and Omori, 1977). It is not known whether they are more susceptible or less susceptible than other ruminant species or whether the predisposing factors are different. For the purpose of this discussion it will be assumed that goats do not differ from cattle or sheep with respect to calculus formation. Nutritional imbalances are generally considered the primary cause of stone formation, but infection has been identified as a predisposing factor with some species (Griffith *et al.*, 1975). The problem is largely restricted to the male because his urinary tract is much more susceptible to blockage, and it is seen infrequently in grazing goats. The problem is important only in confined animals, which represent a small portion of the world's goat population but include some of the more valuable stud bucks. The chemistry of calculus formation is complex and is not completely understood. One of the more important predisposing factors is a high phosphorus content in the diet, or a content high relative to calcium or potassium content (Robbins *et al.*, 1965; Hoar *et al.*, 1970). In dry lot rations the potassium levels should be maintained at an adequate level and the calcium-to-phosphorus ratio should be maintained at 1.5:1 or greater. Additional protection may be obtained through the use of ration additives such as ammonium chloride (Crookshank, 1970) or potassium chloride (Shelton and Ellis, 1965; Crookshank, 1966), which will acidify the urine. If infection plays a part in calculus formation, it may be through its effect on pH of the urine. Using medication to combat infection with a view to preventing calculus formation is not generally recommended, but such an action may be a secondary benefit of using antibiotics in the ration for other reasons.

7 FEED COMPOSITION TABLES

Tables 2, 3, and 4 present the composition of feed ingredients. Nutrient concentrations are organized as follows:

Table 2: Dry matter; total digestible nutrients; digestible, metabolizable, and net energy; crude protein; digestible protein; plant cell wall constituents; and crude fiber
Table 3: Dry matter, minerals, and carotene content
Table 4: Composition of mineral supplements

In Tables 2, 3, and 4 animal feed names follow international nomenclature designed to give a qualitative description of each product, where such information is available and pertinent (NRC publications No. 1684 and No. 1919; Harris *et al.*, 1980). Each feed description is followed by a 5-digit "International Feed Number" (IFN) for identification. A feed-class number placed in front of the international feed number identifies the class to which the feed has been assigned:

Class 1: Dry forages and roughages
This class includes all forages and roughages cut and cured, and other products with more than 18% crude fiber or containing more than 35% cell wall constituents (dry basis). Forages and roughages are usually low in net energy per unit weight because of the high cell wall content. Examples of dry forages and roughages are: hay, straw, fodder (aerial part with ears and husks for the corn plant, or aerial part with heads for the sorghum plant), stover (aerial part without ears or husks for the corn plant, or aerial part without heads for the sorghum plant), hulls, and pods.

Class 2: Pasture, range plants, and forages fed green
This class includes forages on the stem or cut and fed fresh (grasses, shrubs, tree leaves, browse, and forbs).

Class 3: Silages
This class includes ensiled forages (corn, alfalfa, grass, etc.)

Class 4: Energy feeds
This group includes products with less than 20% protein, and less than 18% crude fiber or less than 35% cell wall constituents. Included are grain; mill by-products; fruits; nuts; roots; and tubers, either fresh, dry, or ensiled.

Class 5: Protein supplements
This class includes products which contain 20% or more of protein from plant or animal origins.

Class 6: Mineral supplements

The feed names are listed by the scientific name; however, several feeds are listed by common name since they do not have scientific names: mountain meadow plants (Class 2); animal tallow, bleached, stabilized (Class 4); blood meal, spray dehydrated (Class 5); and meat, with bone meal rendered (Class 5). Table 5 gives an alphabetical list, under five feed classifications, of common and scientific names of the feeds.

Analytical data in Tables 2, 3, and 4 are expressed in the metric system and are on an as-fed and dry basis. Analytical data may differ in various NRC reports because the data are updated for each report. Individual feed samples may vary widely from averages in the table because of factors such as variety, climate, soil, and length of storage. The values given should be used with judgment and related, if possible, to analyses of critical nutrients for the feed on hand. It has not been possible in all cases to obtain data for goats on energy and protein values of feeds; therefore, sheep and cattle data have been used to fill in some information. In Tables 2 and 3, data obtained for goats are not marked. Sheep data are marked with a superior italic a ([a]). Values marked with superior italic ([b]) have been calculated from data for cows by using formulas shown below (Garrett 1976; Moe and Tyrrell, 1976):

$$\% \text{ TDN} = \% \text{ DOM} \div 1.05$$
$$\text{DE Mcal/kg} = \% \text{ TDN} \times 0.04409$$
$$\text{ME Mcal/kg} = 0.82 \times \text{DE Mcal/kg}$$
$$\text{NE}_m \text{ Mcal/kg} = 1.115 - 0.8971 \text{ ME} + 0.6507 \text{ ME}^2 - 0.1028 \text{ ME}^3 + 0.005725 \text{ ME}^4$$
$$\text{NE}_g \text{ Mcal/kg} = 3.178 \text{ ME} - 0.8646 \text{ ME}^2 + 0.1275 \text{ ME}^3 - 0.006787 \text{ ME}^4 - 3.325$$

NE_l Mcal/kg = (0.0245 × % TDN) − 0.12
DP = (0.933 × crude protein) − 3.44 (Class 1 and 2 feeds)
DP = (0.908 × crude protein) − 3.77 (Class 3 feeds)
DP = (0.916 × crude protein) − 2.76 (Class 4 feeds)

International standards for vitamin A activity as related to vitamin A and beta-carotene are as follows:

1 IU vitamin A = 1 USP unit
= vitamin A activity of 0.300 microgram crystalline vitamin A alcohol
= vitamin A activity of 0.344 microgram vitamin A acetate
= vitamin A activity of 0.550 microgram vitamin A palmitate
1 IU vitamin A = 0.6 microgram beta-carotene
1 mg beta-carotene = 1667 IU vitamin A.

International standards for vitamin A are based on the utilization of vitamin A and beta-carotene by the rat. Since goats do not convert carotene to vitamin A in the same ratio as rats, it is suggested that the values in Table 3, when used in connection with the requirements in Table 1, be converted as follows:

1 mg carotene = 400 IU vitamin A.

TABLE 2 Composition of Goat Feeds: Dry Matter; Total Digestible Nutrients; Digestible, Metabolizable, and Net Energy; Crude Protein and Digestible Protein; Plant Cell Wall Constituents; and Acid Detergent Fiber and Crude Fiber; Data Expressed As-Fed and Dry (100% Dry Matter)

Entry Number	SCIENTIFIC NAME Feed Name	International Feed Number	Dry Matter (%)	Feed Energy TDN (%)	DE (Mcal/kg)	ME (Mcal/kg)	NE$_m$ (Mcal/kg)	NE$_g$ (Mcal/kg)	NE$_l$ (Mcal/kg)	Protein Total (%)	Dig. (%)	Cell Walls (%)	Acid Detergent Fiber (%)	Crude Fiber (%)
	DRY FORAGES AND ROUGHAGES													
	Agropyron desertorum. Wheatgrass, crested													
001	hay, sun-cured	1-05-418	93.0	49.0[a]	2.17[b]	1.78[b]	1.06[b]	0.37[b]	1.09[b]	11.5	7.5[b]	—	34.0	30.5
002			100.0	53.0[a]	2.34[b]	1.92[b]	1.14[b]	0.40[b]	1.18[b]	12.4	8.1[b]	—	36.0	32.9
	Agrostis alba. Redtop													
003	hay, sun-cured, full bloom	1-03-882	91.0	51.0[a]	2.24[b]	1.84[b]	1.10[b]	0.46[b]	1.14[b]	8.5	4.8[b]	—	—	28.2
004			100.0	56.0[a]	2.47[b]	2.03[b]	1.21[b]	0.51[b]	1.25[b]	9.4	5.3[b]	—	—	31.1
	Andropogon spp. Bluestem													
005	hay, sun-cured	1-00-819	90.0	41.0[a]	1.78[b]	1.46[b]	0.88[b]	0.05[b]	0.88[b]	4.9	1.4[b]	—	—	30.8
006			100.0	45.0[a]	1.98[b]	1.63[b]	0.98[b]	0.06[b]	0.98[b]	5.4	1.6[b]	—	—	34.3
	Arachis hypogaea. Peanut													
007	hay, sun-cured	1-03-619	91.0	50.0[a]	2.20[b]	1.81[b]	1.08[b]	0.43[b]	1.12[b]	9.8	6.0[b]	—	—	30.2
008			100.0	55.0[a]	2.43[b]	1.99[b]	1.19[b]	0.47[b]	1.23[b]	10.8	6.6[b]	—	—	33.2
009	hay with fruit, sun-cured	1-03-620	92.0	71.0[a]	3.15[b]	2.58[b]	1.66[b]	1.09[b]	1.64[b]	13.3	9.3[b]	—	—	22.8
010			100.0	78.0[a]	3.44[b]	2.82[b]	1.82[b]	1.19[b]	1.79[b]	14.5	10.1[b]	—	—	24.8
011	pods	1-08-028	91.0	20.0[a]	0.89[b]	0.73[b]	0.70[b]	0.00[b]	0.38[b]	7.1	3.5[b]	67.0	59.0	57.3
012			100.0	22.0[a]	0.97[b]	0.80[b]	0.76[b]	0.00[b]	0.42[b]	7.8	3.9[b]	74.0	65.0	62.9
013	straw	1-30-121	86.0	21.0	0.91[b]	0.75[b]	0.66[b]	0.00[b]	0.40[b]	5.7	0.8	—	—	57.2
014			100.0	24.0	1.06[b]	0.87[b]	0.76[b]	0.00[b]	0.47[b]	6.6	0.9	—	—	66.3
	Avena sativa. Oats													
015	hay, sun-cured	1-03-280	91.0	56.0[a]	2.46[b]	2.01[b]	1.22[b]	0.63[b]	1.26[b]	8.5	4.8[b]	60.0	33.0	27.8
016			100.0	61.0[a]	2.69[b]	2.21[b]	1.33[b]	0.69[b]	1.38[b]	9.3	5.3[b]	66.0	36.0	30.4
017	straw	1-03-283	92.0	46.0[a]	2.03[b]	1.67[b]	0.99[b]	0.25[b]	1.02[b]	4.1	0.6[b]	64.0	43.0	37.3
018			100.0	50.0[a]	2.21[b]	1.81[b]	1.07[b]	0.28[b]	1.11[b]	4.4	0.7[b]	70.0	47.0	40.5
	Brachiaria mutica. Paragrass													
019	hay, sun-cured	1-03-517	90.0	39.0[a]	1.70[b]	1.39[b]	0.84[b]	0.00[b]	0.84[b]	3.9	0.5[b]	—	—	31.9
020			100.0	43.0[a]	1.90[b]	1.56[b]	0.94[b]	0.00[b]	0.93[b]	4.3	0.6[b]	—	—	35.6
	Bromus inermis. Brome, smooth													
021	hay, sun-cured, early bloom	1-00-941	91.0	51.0[a]	2.25[b]	1.85[b]	1.10[b]	0.46[b]	1.14[b]	11.0	7.2[b]	56.0	31.0	28.4
022			100.0	56.0[a]	2.47[b]	2.03[b]	1.21[b]	0.51[b]	1.25[b]	12.1	7.8[b]	62.0	34.0	31.2
023	hay, sun-cured, mature	1-00-944	93.0	48.0[a]	2.12[b]	1.74[b]	1.03[b]	0.33[b]	1.07[b]	5.4	1.8[b]	65.0	42.0	29.8
024			100.0	52.0[a]	2.29[b]	1.88[b]	1.12[b]	0.36[b]	1.15[b]	5.8	2.0[b]	71.0	45.0	32.2
	Cajanus cajan. Pigeonpea													
025	straw	1-30-108	89.0	44.0	1.96[b]	1.61[b]	0.96[b]	0.25[b]	0.98[b]	9.5	3.9	—	—	25.5
026			100.0	50.0	2.21[b]	1.81[b]	1.07[b]	0.28[b]	1.11[b]	10.7	4.4	—	—	28.7
	Carthamus tinctorius. Safflower													
027	hay, sun-cured	1-30-127	85.0	58.0[a]	2.56[b]	2.10[b]	1.30[b]	0.78[b]	1.32[b]	10.9	9.0[a]	—	—	27.0
028			100.0	68.0[a]	3.00[b]	2.46[b]	1.52[b]	0.91[b]	1.55[b]	12.8	10.5[a]	—	—	31.7
	Cicer arietinum. Chickpea													
029	pods, sun-cured	1-27-149	95.0	47.0	2.05[b]	1.68[b]	1.00[b]	0.22[b]	1.03[b]	5.4	2.5	—	—	38.8
030			100.0	49.0	2.16[b]	1.77[b]	1.05[b]	0.23[b]	1.08[b]	5.7	2.6	—	—	40.8
031	straw	1-01-217	91.0	45.0	2.00[b]	1.64[b]	0.98[b]	0.25[b]	1.00[b]	4.8	0.8	—	—	38.8
032			100.0	50.0	2.21[b]	1.81[b]	1.07[b]	0.28[b]	1.11[b]	5.3	0.8	—	—	42.7

TABLE 2 Composition of Goat Feeds—*Continued*

Entry Number	SCIENTIFIC NAME Feed Name	International Feed Number	Dry Matter (%)	Feed Energy TDN (%)	DE (Mcal/kg)	ME (Mcal/kg)	NE_m (Mcal/kg)	NE_g (Mcal/kg)	NE_l (Mcal/kg)	Protein Total (%)	Dig. (%)	Cell Walls (%)	Acid Detergent Fiber (%)	Crude Fiber (%)
	Chloris gayana. Rhodesgrass													
033	hay, sun-cured, early	1-03-910	88.0	52.0[a]	2.29[b]	1.88[b]	1.13[b]	0.54[b]	1.17[b]	4.8	1.5[b]	—	—	—
034	vegetative		100.0	59.0[a]	2.60[b]	2.13[b]	1.28[b]	0.62[b]	1.33[b]	5.5	1.7[b]	—	—	—
035	hay, sun-cured, mature	1-03-911	93.0	51.0[a]	2.25[b]	1.85[b]	1.10[b]	0.44[b]	1.14[b]	5.5	1.9[b]	—	—	32.1
036			100.0	55.0[a]	2.43[b]	1.99[b]	1.19[b]	0.47[b]	1.23[b]	5.9	2.1[b]	—	—	34.6
	Cynodon dactylon. Bermudagrass													
037	hay, sun-cured	1-00-703	91.0	42.0	1.85[b]	1.52[b]	0.91[b]	0.10[b]	0.92[b]	8.9	4.2	—	—	27.8
038			100.0	46.0	2.03[b]	1.66[b]	0.99[b]	0.10[b]	1.01[b]	9.8	4.6	—	—	30.4
	Cynodon dactylon. Bermudagrass, coastal													
039	hay, sun-cured, early	1-00-713	94.0	57.0[a]	2.53[b]	2.07[b]	1.25[b]	0.64[b]	1.29[b]	15.0	10.8[b]	62.0	29.0	25.2
040	vegetative		100.0	61.0[a]	2.69[b]	2.21[b]	1.33[b]	0.69[b]	1.38[b]	16.0	11.5[b]	66.0	30.0	26.8
041	hay, sun-cured, late vege-	1-20-900	91.0	50.0[a]	2.17[b]	1.78[b]	1.06[b]	0.40[b]	1.10[b]	15.0	10.9[b]	—	—	24.8
042	tative		100.0	54.0[a]	2.38[b]	1.95[b]	1.16[b]	0.43[b]	1.20[b]	16.5	12.0[b]	—	—	27.3
	Dactylis glomerata. Orchardgrass													
043	hay, sun-cured, cut 1	1-03-433	92.0	55.0[a]	2.44[b]	2.00[b]	1.21[b]	0.60[b]	1.25[b]	13.9	9.8[b]	57.0	34.0	—
044			100.0	60.0[a]	2.65[b]	2.17[b]	1.31[b]	0.65[b]	1.35[b]	15.1	10.6[b]	62.0	37.0	—
045	hay, sun-cured, cut 2	1-03-434	89.0	57.0[a]	2.55[b]	2.09[b]	1.28[b]	0.73[b]	1.31[b]	11.3	7.5[b]	53.0	33.0	31.7
046			100.0	65.0[a]	2.87[b]	2.35[b]	1.44[b]	0.82[b]	1.47[b]	12.7	8.4[b]	60.0	37.0	35.6
	Digitaria decumbens, Pangolagrass													
047	hay, sun-cured, milk stage	1-03-490	88.0	45.0[a]	2.01[b]	1.65[b]	0.98[b]	0.31[b]	1.01[b]	8.4	4.8[b]	—	—	24.0
048			100.0	52.0[a]	2.29[b]	1.88[b]	1.12[b]	0.36[b]	1.15[b]	9.6	5.5[b]	—	—	27.4
	Echinochloa crusgalli. Barnyardgrass													
049	hay, sun-cured	1-03-105	88.0	46.0[a]	2.02[b]	1.65[b]	0.98[b]	0.31[b]	1.02[b]	7.7	4.2[b]	—	—	27.5
050			100.0	52.0[a]	2.29[b]	1.88[b]	1.12[b]	0.36[b]	1.15[b]	8.8	4.8[b]	—	—	31.3
	Eragrostis spp. Lovegrass													
051	hay, sun-cured, full bloom	1-02-644	92.0	51.0[a]	2.27[b]	1.86[b]	1.11[b]	0.47[b]	1.15[b]	7.8	4.1[b]	—	—	—
052			100.0	56.0[a]	2.47[b]	2.03[b]	1.21[b]	0.51[b]	1.25[b]	8.5	4.5[b]	—	—	—
	Festuca arundinacea. Fescue, Kentucky 31													
053	hay, sun-cured, early	1-09-186	91.0	56.0[a]	2.49[b]	2.04[b]	1.24[b]	0.65[b]	1.27[b]	18.4	14.1[b]	—	—	21.5
054	bloom		100.0	62.0[a]	2.73[b]	2.24[b]	1.36[b]	0.72[b]	1.40[b]	20.2	15.4[b]	—	—	23.6
	Glycine max. Soybean													
055	hay, sun-cured, midbloom	1-04-538	94.0	50.0[a]	2.19[b]	1.79[b]	1.07[b]	0.37[b]	1.10[b]	16.7	12.3[b]	—	—	27.9
056			100.0	53.0[a]	2.34[b]	1.92[b]	1.14[b]	0.40[b]	1.18[b]	17.8	13.2[b]	—	—	29.8
057	hay, sun-cured, mature	1-04-543	90.0	49.0[a]	2.15[b]	1.77[b]	1.05[b]	0.39[b]	1.09[b]	13.0	9.0[b]	—	—	32.4
058			100.0	54.0[a]	2.38[b]	1.95[b]	1.16[b]	0.43[b]	1.20[b]	14.4	10.0[b]	—	—	35.8
	Gossypium spp. Cotton													
059	gin by-product	1-08-413	90.0	42.0[a]	1.83[b]	1.50[b]	0.90[b]	0.09[b]	0.91[b]	6.7	3.1[b]	—	—	30.5
060			100.0	46.0[a]	2.03[b]	1.66[b]	0.99[b]	0.10[b]	1.01[b]	7.4	3.5[b]	—	—	33.8
061	hulls	1-01-599	91.0	41.0[a]	1.80[b]	1.47[b]	0.88[b]	0.05[b]	0.89[b]	3.7	0.3[b]	82.0	66.0	43.3
062			100.0	45.0[a]	1.98[b]	1.63[b]	0.98[b]	0.06[b]	0.98[b]	4.1	0.4[b]	90.0	73.0	47.8
	Hordeum vulgare. Barley													
063	hay, sun-cured, milk stage	1-00-490	84.0	43.0[a]	1.90[b]	1.55[b]	0.92[b]	0.27[b]	0.95[b]	7.1	3.7[b]	—	—	27.3

TABLE 2 Composition of Goat Feeds—*Continued*

Entry Number	SCIENTIFIC NAME Feed Name	International Feed Number	Dry Matter (%)	Feed Energy TDN (%)	DE (Mcal/kg)	ME (Mcal/kg)	NE_m (Mcal/kg)	NE_g (Mcal/kg)	NE_l (Mcal/kg)	Protein Total (%)	Dig. (%)	Cell Walls (%)	Acid Detergent Fiber (%)	Crude Fiber (%)
064			100.0	51.0[a]	2.25[b]	1.84[b]	1.10[b]	0.32[b]	1.13[b]	8.4	4.4[b]	—	—	32.4
065	straw	1-00-498	91.0	45.0[a]	1.97[b]	1.61[b]	0.96[b]	0.21[b]	0.99[b]	4.0	0.6[b]	73.0	54.0	38.3
066			100.0	49.0[a]	2.16[b]	1.77[b]	1.05[b]	0.23[b]	1.08[b]	4.3	0.6[b]	80.0	59.0	42.0
	Kochia spp. Summer cypress													
067	hay, sun-cured	1-04-713	92.0	50.0[a]	2.23[b]	1.83[b]	1.09[b]	0.43[b]	1.13[b]	12.6	8.6[b]	—	—	23.6
068			100.0	55.0[a]	2.43[b]	1.99[b]	1.19[b]	0.47[b]	1.23[b]	13.7	9.3[b]	—	—	25.7
	Lespedeza striata. Lespedeza, common													
069	hay, sun-cured, early bloom	1-20-882	91.0	54.0[a]	2.37[b]	1.94[b]	1.17[b]	0.56[b]	1.21[b]	13.7	9.6[b]	—	—	—
070			100.0	59.0[a]	2.60[b]	2.13[b]	1.28[b]	0.62[b]	1.33[b]	15.0	10.6[b]	—	—	—
071	hay, sun-cured, full bloom	1-20-887	89.0	52.0[a]	2.28[b]	1.87[b]	1.12[b]	0.52[b]	1.16[b]	12.8	8.8[b]	—	—	27.4
072			100.0	58.0[a]	2.56[b]	2.10[b]	1.26[b]	0.58[b]	1.30[b]	14.3	9.9[b]	—	—	30.7
	Lolium multiflorum. Ryegrass, Italian													
073	hay, sun-cured, early	1-04-064	89.0	61.0[a]	2.68[b]	2.20[b]	1.36[b]	0.81[b]	1.38[b]	13.6	9.6[b]	—	—	17.6
074	vegetative		100.0	68.0[a]	3.00[b]	2.46[b]	1.52[b]	0.91[b]	1.55[b]	15.2	10.7[b]	—	—	19.7
075	hay, sun-cured, late	1-04-065	86.0	53.0[a]	2.34[b]	1.92[b]	1.16[b]	0.62[b]	1.20[b]	8.8	5.3[b]	—	—	20.4
076	vegetative		100.0	62.0[a]	2.73[b]	2.24[b]	1.36[b]	0.72[b]	1.40[b]	10.3	6.2[b]	—	—	23.8
	Lotus corniculatus. Trefoil, birdsfoot													
077	hay, sun-cured, midbloom	1-20-790	91.0	57.0[a]	2.53[b]	2.07[b]	1.26[b]	0.68[b]	1.30[b]	13.2	9.2[b]	45.0	34.0	—
078			100.0	63.0[a]	2.78[b]	2.28[b]	1.39[b]	0.75[b]	1.42[b]	14.5	10.1[b]	50.0	38.0	—
	Medicago sativa. Alfalfa													
079	meal, 17% protein	1-00-023	92.0	56.0[a]	2.47[b]	2.02[b]	1.22[b]	0.63[b]	1.26[b]	17.3	13.0[b]	41.0	32.0	24.0
080			100.0	61.0[a]	2.69[b]	2.21[b]	1.33[b]	0.69[b]	1.38[b]	18.9	14.2[b]	45.0	35.0	26.2
081	hay, sun-cured, early	1-00-050	90.0	59.0[a]	2.62[b]	2.15[b]	1.32[b]	0.76[b]	1.35[b]	20.7	16.2[b]	34.0	23.0	18.4
082	vegetative		100.0	66.0[a]	2.91[b]	2.39[b]	1.47[b]	0.85[b]	1.50[b]	23.0	18.0[b]	38.0	26.0	20.5
083	hay, sun-cured, late	1-00-054	90.0	57.0[a]	2.49[b]	2.04[b]	1.24[b]	0.67[b]	1.28[b]	17.9	13.7[b]	36.0	25.0	22.4
084	vegetative		100.0	63.0[a]	2.78[b]	2.28[b]	1.39[b]	0.75[b]	1.42[b]	20.0	15.2[b]	40.0	28.0	25.0
085	hay, sun-cured, early bloom	1-00-059	90.0	54.0[a]	2.38[b]	1.95[b]	1.18[b]	0.59[b]	1.22[b]	16.2	12.0[b]	38.0	27.0	23.4
086			100.0	60.0[a]	2.65[b]	2.17[b]	1.31[b]	0.65[b]	1.35[b]	18.0	13.4[b]	42.0	30.0	26.0
087	hay, sun-cured, full bloom	1-00-068	90.0	50.0[a]	2.18[b]	1.79[b]	1.07[b]	0.43[b]	1.11[b]	13.5	9.5[b]	45.0	32.0	26.1
088			100.0	55.0[a]	2.43[b]	1.99[b]	1.19[b]	0.47[b]	1.23[b]	15.0	10.6[b]	50.0	36.0	29.0
089	hay, sun-cured, mature	1-00-071	91.0	49.0[a]	2.17[b]	1.78[b]	1.06[b]	0.40[b]	1.10[b]	11.7	7.8[b]	53.0	40.0	34.4
090			100.0	54.0[a]	2.38[b]	1.95[b]	1.16[b]	0.43[b]	1.20[b]	12.9	8.6[b]	58.0	44.0	37.7
	Oryza sativa. Rice													
091	straw	1-03-925	91.0	41.0	1.80[b]	1.47[b]	0.88[b]	0.05[b]	0.89[b]	4.0	0.6	—	—	32.0
092			100.0	45.0	1.98[b]	1.63[b]	0.98[b]	0.06[b]	0.98[b]	4.4	0.6	—	—	35.3
	Panicum maximum. Guineagrass													
093	dehydrated pelleted, 15	1-30-084	92.0	53.0[a]	2.36[b]	1.93[b]	1.16[b]	0.54[b]	1.20[b]	15.4	—	—	—	26.0
094	to 28 days' growth		100.0	58.0[a]	2.56[b]	2.10[b]	1.26[b]	0.58[b]	1.30[b]	16.7	—	—	—	28.2
095	hay, sun-cured	1-02-336	89.0	45.0[a]	1.97[b]	1.61[b]	0.96[b]	0.25[b]	0.99[b]	6.3	2.8[b]	—	—	28.1
096			100.0	50.0[a]	2.21[b]	1.81[b]	1.07[b]	0.28[b]	1.11[b]	7.0	3.1[b]	—	—	31.5
	Paspalum dilatatum. Dallisgrass													
097	hay, sun-cured, midbloom	1-01-734	91.0	47.0[a]	2.04[b]	1.67[b]	0.99[b]	0.29[b]	1.02[b]	6.5	3.0[b]	—	—	29.1
098			100.0	51.0[a]	2.25[b]	1.84[b]	1.10[b]	0.32[b]	1.13[b]	7.2	3.3[b]	—	—	32.1

TABLE 2 Composition of Goat Feeds—*Continued*

Entry Number	SCIENTIFIC NAME Feed Name	International Feed Number	Dry Matter (%)	Feed Energy TDN (%)	DE (Mcal/kg)	ME (Mcal/kg)	NE_m (Mcal/kg)	NE_g (Mcal/kg)	NE_l (Mcal/kg)	Protein Total (%)	Protein Dig. (%)	Cell Walls (%)	Acid Detergent Fiber (%)	Crude Fiber (%)
	Paspalum notatum. Bahiagrass													
099	hay, sun-cured	1-00-462	91.0	47.0[a]	2.05[b]	1.69[b]	1.00[b]	0.29[b]	1.03[b]	7.4	3.8[b]	66.0	37.0	29.2
100			100.0	51.0[a]	2.25[b]	1.84[b]	1.10[b]	0.32[b]	1.13[b]	8.2	4.2[b]	72.0	41.0	32.0
	Pennisetum purpureum. Napiergrass													
101	hay, sun-cured, dough stage	1-20-809	91.0	45.0[a]	1.96[b]	1.61[b]	0.96[b]	0.21[b]	0.98[b]	7.7	4.1[b]	59.0	40.0	30.5
102			100.0	49.0[a]	2.16[b]	1.77[b]	1.05[b]	0.23[b]	1.08[b]	8.5	4.5[b]	65.0	44.0	33.6
103	hay, sun-cured, mature	1-20-808	90.0	42.0[a]	1.87[b]	1.53[b]	0.91[b]	0.13[b]	0.93[b]	4.9	1.4[b]	58.0	42.0	29.4
104			100.0	47.0[a]	2.07[b]	1.70[b]	1.01[b]	0.15[b]	1.03[b]	5.4	1.6[b]	65.0	46.0	32.6
	Phalaris arundinacea. Canarygrass, reed													
105	hay, sun-cured	1-01-104	91.0	50.0[a]	2.21[b]	1.81[b]	1.08[b]	0.43[b]	1.12[b]	9.4	5.6[b]	58.0	33.0	30.1
106			100.0	55.0[a]	2.43[b]	1.99[b]	1.19[b]	0.47[b]	1.23[b]	10.3	6.2[b]	64.0	36.0	33.0
	Phleum pratense. Timothy													
107	hay, sun-cured, late	1-04-881	89.0	57.0[a]	2.52[b]	2.07[b]	1.26[b]	0.70[b]	1.29[b]	10.4	6.6[b]	49.0	27.0	24.1
108	vegetative		100.0	64.0[a]	2.82[b]	2.31[b]	1.41[b]	0.78[b]	1.45[b]	11.7	7.4[b]	55.0	30.0	27.0
109	hay, sun-cured, full bloom	1-04-884	89.0	51.0[a]	2.27[b]	1.86[b]	1.12[b]	0.52[b]	1.15[b]	6.8	3.3[b]	—	—	31.4
110			100.0	58.0[a]	2.56[b]	2.10[b]	1.26[b]	0.58[b]	1.30[b]	7.6	3.7[b]	—	—	35.4
	Pisum sativum sativum. Pea, garden													
111	pods, fresh	1-23-550	18.0	13.0[a]	0.56[b]	0.46[b]	0.29[b]	0.17[b]	0.29[b]	2.7	1.9	—	—	—
112			100.0	70.0[a]	3.09[b]	2.53[b]	1.58[b]	0.97[b]	1.60[b]	14.8	10.4	—	—	—
	Poa pratensis. Bluegrass, Kentucky													
113	hay, sun-cured	1-00-776	89.0	54.0[a]	2.40[b]	1.96[b]	1.19[b]	0.61[b]	1.22[b]	8.4	4.8[b]	—	—	27.6
114			100.0	61.0[a]	2.69[b]	2.21[b]	1.33[b]	0.69[b]	1.38[b]	9.4	5.4[b]	—	—	31.0
	Prosopis spicigera. Mesquite, spicigera													
115	hay, sun-cured	1-30-160	86.0	35.0	1.55[b]	1.27[b]	0.78[b]	0.00[b]	0.76[b]	12.2	3.9	—	—	18.9
116			100.0	41.0	1.81[b]	1.48[b]	0.91[b]	0.00[b]	0.89[b]	14.2	4.5	—	—	22.1
	Pueraria spp. Kudzu													
117	hay, sun-cured	1-02-478	91.0	50.0[a]	2.22[b]	1.82[b]	1.08[b]	0.43[b]	1.12[b]	13.1	9.1[b]	—	—	35.7
118			100.0	55.0[a]	2.43[b]	1.99[b]	1.19[b]	0.47[b]	1.23[b]	14.3	9.9[b]	—	—	39.1
	Saccharum officinarum. Sugarcane													
119	bagasse, dehydrated	1-04-686	91.0	44.0[a]	1.93[b]	1.59[b]	0.94[b]	0.17[b]	0.96[b]	1.5	0.0[b]	—	—	43.9
120			100.0	48.0[a]	2.12[b]	1.74[b]	1.03[b]	0.19[b]	1.06[b]	1.6	0.0[b]	—	—	48.1
	Secale cereale. Rye													
121	hay, sun-cured	1-04-004	93.0	43.0[a]	1.94[b]	1.59[b]	0.95[b]	0.14[b]	0.96[b]	7.9	4.2[b]	—	—	31.1
122			100.0	47.0[a]	2.07[b]	1.70[b]	1.01[b]	0.15[b]	1.03[b]	8.5	4.5[b]	—	—	33.3
123	straw	1-04-007	90.0	40.0[a]	1.74[b]	1.43[b]	0.86[b]	0.01[b]	0.86[b]	2.7	0.0[b]	—	—	38.7
124			100.0	44.0[a]	1.94[b]	1.59[b]	0.96[b]	0.01[b]	0.96[b]	3.0	0.0[b]	—	—	43.1
	Sorghum bicolor. Sorghum													
125	hay, sun-cured, full bloom	1-04-371	90.0	59.0	2.60[b]	2.14[b]	1.31[b]	0.76[b]	1.34[b]	4.7	2.8	—	—	25.9
126			100.0	66.0	2.91[b]	2.39[b]	1.47[b]	0.85[b]	1.50[b]	5.2	3.1	—	—	28.9
127	aerial part, sun-cured,	1-04-301	90.0	48.0[a]	2.11[b]	1.73[b]	1.03[b]	0.36[b]	1.06[b]	6.1	2.5[b]	—	—	25.2
128	mature		100.0	53.0[a]	2.34[b]	1.92[b]	1.14[b]	0.40[b]	1.18[b]	6.7	2.8[b]	—	—	28.0

TABLE 2 Composition of Goat Feeds—*Continued*

Entry Number	SCIENTIFIC NAME Feed Name	International Feed Number	Dry Matter (%)	Feed Energy TDN (%)	DE (Mcal/kg)	ME (Mcal/kg)	NE_m (Mcal/kg)	NE_g (Mcal/kg)	NE_l (Mcal/kg)	Protein Total (%)	Dig. (%)	Cell Walls (%)	Acid Detergent Fiber (%)	Crude Fiber (%)
	Sorghum bicolor sudanense. Sorghum, Sudangrass													
129	hay, sun-cured, early vegetative	1-04-473	90.0	52.0[a]	2.30[b]	1.88[b]	1.13[b]	0.52[b]	1.17[b]	14.0	10.0[b]	47.0	26.0	23.0
130			100.0	58.0[a]	2.56[b]	2.10[b]	1.26[b]	0.58[b]	1.30[b]	15.6	11.1[b]	53.0	29.0	25.6
131	hay, sun-cured, late vegetative	1-04-474	88.0	50.0[a]	2.21[b]	1.81[b]	1.08[b]	0.48[b]	1.12[b]	12.2	8.3[b]	52.0	29.0	28.0
132			100.0	57.0[a]	2.51[b]	2.06[b]	1.23[b]	0.55[b]	1.28[b]	13.9	9.5[b]	60.0	33.0	31.9
	Sorghum halepense. Sorghum, Johnsongrass													
133	hay, sun-cured	1-04-407	89.0	47.0[a]	2.09[b]	1.71[b]	1.02[b]	0.35[b]	1.05[b]	8.5	4.8[b]	—	—	29.9
134			100.0	53.0[a]	2.34[b]	1.92[b]	1.14[b]	0.40[b]	1.18[b]	9.5	5.4[b]	—	—	33.5
	Trifolium alexandrinum. Clover, Egyptian													
135	hay, sun-cured	1-01-340	88.0	57.0	2.54[b]	2.08[b]	1.28[b]	0.72[b]	1.30[b]	13.4	9.3	—	—	23.0
136			100.0	65.0	2.87[b]	2.35[b]	1.44[b]	0.82[b]	1.47[b]	15.2	10.5	—	—	26.0
	Trifolium pratense. Clover, red													
137	hay, sun-cured	1-01-415	89.0	49.0[a]	2.15[b]	1.76[b]	1.05[b]	0.42[b]	1.09[b]	14.2	10.2[b]	50.0	—	25.5
138			100.0	55.0[a]	2.43[b]	1.99[b]	1.19[b]	0.47[b]	1.23[b]	16.0	11.5[b]	56.0	—	28.8
	Triticum aestivum. Wheat													
139	straw	1-05-175	89.0	39.0	1.72[b]	1.41[b]	0.85[b]	0.01[b]	0.85[b]	3.2	0.0[b]	75.0	48.0	36.9
140			100.0	44.0	1.94[b]	1.59[b]	0.96[b]	0.01[b]	0.96[b]	3.6	0.0[b]	85.0	54.0	41.6
	Vicia spp. Vetch													
141	hay, sun-cured	1-05-106	89.0	51.0[a]	2.24[b]	1.84[b]	1.10[b]	0.49[b]	1.14[b]	18.5	14.2[b]	43.0	30.0	27.3
142			100.0	57.0[a]	2.51[b]	2.06[b]	1.23[b]	0.55[b]	1.28[b]	20.8	15.9[b]	48.0	33.0	30.6
	Vigna sinensis. Cowpea, common													
143	hay, sun-cured	1-01-645	90.0	53.0[a]	2.34[b]	1.92[b]	1.15[b]	0.56[b]	1.19[b]	17.5	13.2[b]	—	—	24.0
144			100.0	59.0[a]	2.60[b]	2.13[b]	1.28[b]	0.62[b]	1.33[b]	19.4	14.7[b]	—	—	26.7
145	hay, sun-cured, mature	1-01-644	90.0	52.0[a]	2.31[b]	1.89[b]	1.13[b]	0.52[b]	1.17[b]	10.2	6.4[b]	—	—	30.7
146			100.0	58.0[a]	2.56[b]	2.10[b]	1.26[b]	0.58[b]	1.30[b]	11.3	7.1[b]	—	—	34.0
	Wood													
147	sawdust	1-07-714	90.0	80.0[a]	3.49[b]	2.86[b]	1.91[b]	1.30[b]	1.83[b]	0.3	0.0[b]	—	—	71.5
148			100.0	88.0[a]	3.88[b]	3.18[b]	2.12[b]	1.45[b]	2.04[b]	0.3	0.0[b]	—	—	79.5
	Zea mays. Corn													
149	cobs, ground	1-02-782	90.0	45.0[a]	1.99[b]	1.63[b]	0.97[b]	0.25[b]	1.00[b]	2.8	0.0[b]	80.0	32.0	32.7
150			100.0	50.0[a]	2.21[b]	1.81[b]	1.07[b]	0.28[b]	1.11[b]	3.2	0.0[b]	89.0	35.0	36.2
	Ziziphus nummularia. Jujube, wild													
151	hay, sun-cured	1-30-170	91.0	28.0[a]	1.24[b]	1.02[b]	0.72[b]	0.00[b]	0.58[b]	13.5	9.4[b]	—	—	14.2
152			100.0	31.0[a]	1.37[b]	1.12[b]	0.79[b]	0.00[b]	0.64[b]	14.8	10.4[b]	—	—	15.6
	PASTURE, RANGE PLANTS, AND FORAGES FED GREEN													
	Acacia catechu. Acacia, catechu													
153	leaves, fresh	2-30-132	30.0	6.0	0.25[b]	0.21[b]	0.23[b]	0.00[b]	0.10[b]	1.5	0.4	—	—	6.8
154			100.0	19.0	0.84[b]	0.69[b]	0.77[b]	0.00[b]	0.35[b]	5.0	1.2	—	—	22.6

TABLE 2 Composition of Goat Feeds—*Continued*

Entry Number	SCIENTIFIC NAME Feed Name	International Feed Number	Dry Matter (%)	Feed Energy TDN (%)	DE (Mcal/kg)	ME (Mcal/kg)	NE_m (Mcal/kg)	NE_g (Mcal/kg)	NE_l (Mcal/kg)	Protein Total (%)	Dig. (%)	Cell Walls (%)	Acid Detergent Fiber (%)	Crude Fiber (%)
	Agave spp. Agave													
155	browse, fresh	2-00-014	12.0	7.0[a]	0.32[b]	0.27[b]	0.16[b]	0.08[b]	0.17[b]	0.8	0.3[b]	—	—	3.3
156			100.0	61.0[a]	2.69[b]	2.21[b]	1.33[b]	0.69[b]	1.38[b]	6.8	2.9[b]	—	—	27.3
	Agropyron desertorum. Wheatgrass, crested													
157	fresh, early vegetative	2-05-420	28.0	21.0[a]	0.92[b]	0.76[b]	0.48[b]	0.31[b]	0.48[b]	6.0	4.7[b]	—	—	6.2
158			100.0	75.0[a]	3.31[b]	2.71[b]	1.73[b]	1.11[b]	1.72[b]	21.5	16.7[b]	—	—	22.2
159	fresh, post ripe	2-05-428	80.0	39.0[a]	1.73[b]	1.42[b]	0.84[b]	0.19[b]	0.86[b]	2.5	0.0[b]	—	—	32.2
160			100.0	49.0[a]	2.16[b]	1.77[b]	1.05[b]	0.23[b]	1.08[b]	3.1	0.0[b]	—	—	40.3
	Agropyron smithii. Wheatgrass, western													
161	fresh, mature	2-05-407	63.0	31.0[a]	1.39[b]	1.14[b]	0.68[b]	0.17[b]	0.70[b]	3.3	0.9[b]	—	—	21.0
162			100.0	50.0[a]	2.21[b]	1.81[b]	1.07[b]	0.28[b]	1.11[b]	5.3	1.5[b]	—	—	33.3
	Agrostis alba. Redtop													
163	fresh, midbloom	2-03-890	39.0	24.0[a]	1.07[b]	0.87[b]	0.53[b]	0.28[b]	0.55[b]	2.9	1.4[b]	—	—	12.6
164			100.0	62.0[a]	2.73[b]	2.24[b]	1.36[b]	0.72[b]	1.40[b]	7.4	3.5[b]	—	—	32.3
165	fresh, mature	2-03-894	51.0	30.0[a]	1.33[b]	1.09[b]	0.65[b]	0.32[b]	0.68[b]	2.9	1.0[b]	—	—	14.5
166			100.0	59.0[a]	2.60[b]	2.13[b]	1.28[b]	0.62[b]	1.33[b]	5.7	1.9[b]	—	—	28.5
	Ailanthus excelsa. Ailanthus, excelsa													
167	browse, fresh	2-30-096	26.0	17.0	0.73[b]	0.60[b]	0.37[b]	0.20[b]	0.38[b]	4.2	3.4	—	—	5.7
168			100.0	64.0	2.82[b]	2.31[b]	1.41[b]	0.78[b]	1.45[b]	16.3	13.0	—	—	21.9
	Albizia lebbeck. Lebbeck tree													
169	leaves, fresh	2-25-615	17.0	8.0	0.36[b]	0.29[b]	0.17[b]	0.04[b]	0.18[b]	2.8	1.9	—	—	5.2
170			100.0	49.0	2.16[b]	1.77[b]	1.05[b]	0.23[b]	1.08[b]	16.8	11.6	—	—	31.5
	Ambrosia spp. Ragweed													
171	fresh, early vegetative	2-03-851	27.0	18.0[a]	0.79[b]	0.64[b]	0.40[b]	0.23[b]	0.41[b]	4.4	3.2[b]	—	—	5.3
172			100.0	67.0[a]	2.95[b]	2.42[b]	1.50[b]	0.88[b]	1.52[b]	16.7	12.1[b]	—	—	19.8
	Andropogon barbinodis. Bluestem, cane													
173	fresh, late vegetative	2-30-195	46.0	22.0[b]	0.97[b]	0.79[b]	0.47[b]	0.09[b]	0.48[b]	3.0	1.2[b]	32.0	—	—
174			100.0	48.0[b]	2.12[b]	1.74[b]	1.03[b]	0.19[b]	1.06[b]	6.5	2.6[b]	70.0	—	—
175	fresh, mature	2-30-057	89.0	41.0[b]	1.81[b]	1.48[b]	0.88[b]	0.09[b]	0.90[b]	2.7	0.0[b]	63.0	—	—
176			100.0	46.0[b]	2.03[b]	1.66[b]	0.99[b]	0.10[b]	1.01[b]	3.0	0.0[b]	71.0	—	—
	Andropogon scoparius. Bluestem, little													
177	fresh, late vegetative	2-29-994	72.0	40.0[b]	1.74[b]	1.43[b]	0.85[b]	0.34[b]	0.88[b]	6.1	3.2[b]	49.0	—	—
178			100.0	55.0[b]	2.43[b]	1.99[b]	1.19[b]	0.47[b]	1.23[b]	8.5	4.5[b]	68.0	—	—
	Anemone heterophylla. Anemone, heterophylla													
179	fresh, late vegetative	2-30-072	23.0	16.0[b]	0.68[b]	0.56[b]	0.35[b]	0.21[b]	0.35[b]	2.5	1.6[b]	6.0	—	—
180			100.0	68.0[b]	3.00[b]	2.46[b]	1.52[b]	0.91[b]	1.55[b]	11.0	6.8[b]	28.0	—	—
	Aristida oligantha. Threeawn, prairie													
181	fresh, mature	2-04-841	60.0	34.0[a]	1.51[b]	1.24[b]	0.74[b]	0.33[b]	0.77[b]	3.1	0.8[b]	—	—	18.8
182			100.0	57.0[a]	2.51[b]	2.06[b]	1.23[b]	0.55[b]	1.28[b]	5.2	1.4[b]	—	—	31.4

TABLE 2 Composition of Goat Feeds—*Continued*

Entry Number	SCIENTIFIC NAME Feed Name	International Feed Number	Dry Matter (%)	Feed Energy TDN (%)	DE (Mcal/kg)	ME (Mcal/kg)	NE_m (Mcal/kg)	NE_g (Mcal/kg)	NE_l (Mcal/kg)	Protein Total (%)	Protein Dig. (%)	Cell Walls (%)	Acid Detergent Fiber (%)	Crude Fiber (%)
	Aristida wrightii. Threeawn, Wright													
183	fresh, late vegetative	2-30-016	63.0	26.0[b]	1.14[b]	0.93[b]	0.57[b]	0.00[b]	0.56[b]	4.3	1.9[b]	48.0	—	—
184			100.0	41.0[b]	1.81[b]	1.48[b]	0.91[b]	0.00[b]	0.89[b]	6.9	3.0[b]	76.0	—	—
	Artemisia filifolia. Sagebrush, sand													
185	browse, fresh, early	2-04-133	29.0	19.0[a]	0.84[b]	0.69[b]	0.42[b]	0.25[b]	0.43[b]	3.5	2.3[b]	—	—	6.5
186	vegetative		100.0	66.0[a]	2.91[b]	2.39[b]	1.47[b]	0.85[b]	1.50[b]	12.2	7.9[b]	—	—	22.6
187	browse, fresh, mature	2-04-135	45.0	27.0[a]	1.19[b]	0.98[b]	0.59[b]	0.29[b]	0.61[b]	3.2	1.5[b]	—	—	14.3
188			100.0	60.0[a]	2.65[b]	2.17[b]	1.31[b]	0.65[b]	1.35[b]	7.2	3.3[b]	—	—	31.7
	Artemisia ludoviciana albula. Sagebrush, Mexican													
189	browse, fresh, mature	2-30-052	44.0	21.0[b]	0.94[b]	0.77[b]	0.46[b]	0.10[b]	0.47[b]	4.4	2.6[b]	24.0	—	—
190			100.0	49.0[b]	2.16[b]	1.77[b]	1.05[b]	0.23[b]	1.08[b]	10.2	6.1[b]	56.0	—	—
	Aster spp. Aster													
191	fresh	2-00-443	32.0	21.0[a]	0.93[b]	0.76[b]	0.47[b]	0.27[b]	0.48[b]	2.7	1.4[b]	—	—	6.8
192			100.0	66.0[a]	2.91[b]	2.39[b]	1.47[b]	0.85[b]	1.50[b]	8.4	4.4[b]	—	—	21.1
	Astragalus nuttallii. Milkvetch, Nuttall													
193	fresh, early vegetative	2-29-814	28.0	18.0[b]	0.78[b]	0.64[b]	0.39[b]	0.21[b]	0.40[b]	4.2	2.9[b]	8.0	—	—
194			100.0	63.0[b]	2.78[b]	2.28[b]	1.39[b]	0.75[b]	1.42[b]	14.8	10.4[b]	30.0	—	—
	Atriplex confertifolia. Saltbush, shadscale													
195	browse, fresh, early	2-20-233	29.0	15.0[a]	0.65[b]	0.54[b]	0.32[b]	0.09[b]	0.33[b]	2.6	1.4[b]	—	—	8.3
196	vegetative		100.0	51.0[a]	2.25[b]	1.84[b]	1.10[b]	0.32[b]	1.13[b]	9.0	5.0[b]	—	—	28.5
197	browse, fresh, mature	2-20-453	56.0	29.0[a]	1.28[b]	1.05[b]	0.63[b]	0.20[b]	0.65[b]	3.4	1.2[b]	—	—	15.7
198			100.0	52.0[a]	2.29[b]	1.88[b]	1.12[b]	0.36[b]	1.15[b]	6.0	2.2[b]	—	—	28.0
	Avena sativa. Oats													
199	fresh	2-03-292	20.0	12.0[a]	0.55[b]	0.45[b]	0.27[b]	0.14[b]	0.28[b]	2.7	1.9[b]	—	—	6.0
200			100.0	62.0[a]	2.73[b]	2.24[b]	1.36[b]	0.72[b]	1.40[b]	13.6	9.3[b]	—	—	29.7
	Azadirachta indica. Margosa													
201	browse, fresh	2-30-147	21.0	14.0	0.61[b]	0.50[b]	0.31[b]	0.18[b]	0.32[b]	3.9	2.9	—	—	2.5
202			100.0	67.0	2.95[b]	2.42[b]	1.50[b]	0.88[b]	1.52[b]	18.7	13.8	—	—	11.9
203	leaves, fresh	2-27-194	20.0	4.0	0.19[b]	0.15[b]	0.15[b]	0.00[b]	0.08[b]	3.2	0.7	—	—	4.2
204			100.0	21.0	0.93[b]	0.76[b]	0.77[b]	0.00[b]	0.40[b]	16.1	3.4	—	—	20.7
	Bouteloua curtipendula. Grama, sideoats													
205	fresh, late vegetative	2-29-863	62.0	25.0[b]	1.12[b]	0.91[b]	0.56[b]	0.00[b]	0.55[b]	4.1	—	42.0	—	—
206			100.0	41.0[b]	1.81[b]	1.48[b]	0.91[b]	0.00[b]	0.89[b]	6.7	—	67.0	—	—
207	aerial part without heads,	2-30-042	78.0	24.0[b]	1.09[b]	0.90[b]	0.62[b]	0.00[b]	0.52[b]	3.7	—	52.0	—	—
208	fresh, mature		100.0	32.0[b]	1.41[b]	1.16[b]	0.80[b]	0.00[b]	0.66[b]	4.8	—	68.0	—	—
	Bouteloua hirsuta. Grama, hairy													
209	fresh, late vegetative	2-30-006	55.0	27.0[b]	1.21[b]	0.99[b]	0.59[b]	0.15[b]	0.60[b]	3.6	—	39.0	—	—
210			100.0	50.0[b]	2.21[b]	1.81[b]	1.07[b]	0.28[b]	1.11[b]	6.7	—	72.0	—	—

TABLE 2 Composition of Goat Feeds—*Continued*

Entry Number	SCIENTIFIC NAME Feed Name	International Feed Number	Dry Matter (%)	Feed Energy TDN (%)	DE (Mcal/kg)	ME (Mcal/kg)	NE_m (Mcal/kg)	NE_g (Mcal/kg)	NE_l (Mcal/kg)	Protein Total (%)	Dig. (%)	Cell Walls (%)	Acid Detergent Fiber (%)	Crude Fiber (%)
	Brachiaria mutica. Paragrass													
211	fresh	2-03-525	28.0	14.0[a]	0.61[b]	0.50[b]	0.30[b]	0.08[b]	0.31[b]	1.9	0.9[a]	—	—	9.4
212			100.0	50.0[a]	2.21[b]	1.81[b]	1.07[b]	0.28[b]	1.11[b]	7.0	3.3[a]	—	—	33.9
	Brassica campestris. Rape, turnip													
213	fresh	2-03-872	15.0	9.0[a]	0.38[b]	0.31[b]	0.19[b]	0.09[b]	0.19[b]	1.3	0.7[b]	—	—	4.0
214			100.0	58.0[a]	2.56[b]	2.10[b]	1.26[b]	0.58[b]	1.30[b]	8.8	4.8[b]	—	—	26.6
	Brassica oleracea medullosa. Kale, marrow													
215	fresh, 71 to 84 days' growth	2-30-181	11.0	7.0[a]	0.30[b]	0.24[b]	0.15[b]	0.08[b]	0.15[b]	2.3	1.8[b]	—	—	0.9
216			100.0	61.0[a]	2.69[b]	2.21[b]	1.33[b]	0.69[b]	1.38[b]	20.8	16.0[b]	—	—	8.0
217	fresh, 99 to 112 days'	2-30-182	12.0	7.0[a]	0.32[b]	0.27[b]	0.16[b]	0.08[b]	0.17[b]	1.3	0.8[b]	—	—	2.1
218	growth		100.0	60.0[a]	2.65[b]	2.17[b]	1.31[b]	0.65[b]	1.35[b]	10.6	6.5[b]	—	—	17.2
219	leaves, fresh	2-02-457	16.0	12.0[a]	0.55[b]	0.45[b]	0.29[b]	0.19[b]	0.29[b]	2.5	1.8[b]	—	—	1.6
220			100.0	79.0[a]	3.48[b]	2.86[b]	1.85[b]	1.22[b]	1.82[b]	15.8	11.3[b]	—	—	10.5
221	stems, fresh	2-08-449	14.0	10.0[a]	0.44[b]	0.36[b]	0.23[b]	0.14[b]	0.23[b]	1.6	1.0[b]	—	—	3.2
222			100.0	72.0[a]	3.17[b]	2.60[b]	1.64[b]	1.03[b]	1.64[b]	11.4	7.2[b]	—	—	23.2
	Bromus catharticus. Brome, rescue													
223	fresh, late vegetative	2-08-362	33.0	21.0[b]	0.93[b]	0.76[b]	0.46[b]	0.26[b]	0.47[b]	4.4	3.0[b]	19.0	—	10.0
224			100.0	64.0[b]	2.82[b]	2.31[b]	1.41[b]	0.78[b]	1.45[b]	13.6	9.2[b]	57.0	—	30.7
	Buchloe dactyloides. Buffalograss													
225	fresh, late vegetative	2-29-868	61.0	21.0[b]	0.95[b]	0.78[b]	0.51[b]	0.00[b]	0.45[b]	4.7	2.3[b]	41.0	—	—
226			100.0	35.0[b]	1.54[b]	1.27[b]	0.83[b]	0.00[b]	0.74[b]	7.7	3.7[b]	67.0	—	—
227	fresh, mature	2-01-008	72.0	39.0[a]	1.72[b]	1.41[b]	0.84[b]	0.31[b]	0.87[b]	4.3	1.5[b]	—	—	22.2
228			100.0	54.0[a]	2.38[b]	1.95[b]	1.16[b]	0.43[b]	1.20[b]	5.9	2.1[b]	—	—	30.7
	Carex planostachys. Sedge, thicket													
229	fresh, late vegetative	2-29-865	39.0	19.0[b]	0.86[b]	0.71[b]	0.42[b]	0.11[b]	0.43[b]	4.6	3.0[b]	25.0	—	—
230			100.0	50.0[b]	2.21[b]	1.81[b]	1.07[b]	0.28[b]	1.11[b]	11.8	7.6[b]	63.0	—	—
	Carthamus tinctorius. Safflower													
231	fresh	2-28-574	15.0	9.0[a]	0.39[b]	0.32[b]	0.19[b]	0.09[b]	0.20[b]	2.4	1.7[b]	—	—	3.8
232			100.0	58.0[a]	2.56[b]	2.10[b]	1.26[b]	0.58[b]	1.30[b]	15.9	11.4[b]	—	—	25.0
233	leaves, fresh	2-30-129	16.0	10.0[a]	0.43[b]	0.35[b]	0.21[b]	0.11[b]	0.22[b]	4.5	3.7[b]	—	—	1.7
234			100.0	62.0[a]	2.73[b]	2.24[b]	1.36[b]	0.72[b]	1.40[b]	28.9	23.5[b]	—	—	10.9
235	stems, fresh	2-30-130	15.0	6.0[a]	0.28[b]	0.23[b]	0.14[b]	0.00[b]	0.14[b]	1.1	0.5[b]	—	—	6.1
236			100.0	41.0[a]	1.81[b]	1.48[b]	0.91[b]	0.00[b]	0.89[b]	7.4	3.5[b]	—	—	39.8
	Celtis reticulata. Hackberry, netleaf													
237	leaves, fresh	2-30-045	56.0	24.0[b]	1.05[b]	0.86[b]	0.52[b]	0.00[b]	0.52[b]	4.3	2.1[b]	14.0	—	—
238			100.0	43.0[b]	1.90[b]	1.56[b]	0.94[b]	0.00[b]	0.93[b]	7.8	3.8[b]	26.0	—	—
	Celtis tetendra. Hackberry, tetendra													
239	browse, fresh, mature	2-30-164	53.0	21.0[a]	0.95[b]	0.78[b]	0.48[b]	0.00[b]	0.46[b]	7.4	5.1[b]	—	—	8.7
240			100.0	41.0[a]	1.81[b]	1.48[b]	0.91[b]	0.00[b]	0.89[b]	14.0	9.6[b]	—	—	16.6

TABLE 2 Composition of Goat Feeds—*Continued*

Entry Number	SCIENTIFIC NAME Feed Name	International Feed Number	Dry Matter (%)	Feed Energy TDN (%)	DE (Mcal/kg)	ME (Mcal/kg)	NE_m (Mcal/kg)	NE_g (Mcal/kg)	NE_l (Mcal/kg)	Protein Total (%)	Protein Dig. (%)	Cell Walls (%)	Acid Detergent Fiber (%)	Crude Fiber (%)
	Cenchrus ciliaris. Buffelgrass													
241	fresh	2-10-360	32.0	13.0[a]	0.56[b]	0.46[b]	0.29[b]	0.00[b]	0.28[b]	1.8	0.5[b]	—	—	10.8
242			100.0	40.0[a]	1.76[b]	1.45[b]	0.89[b]	0.00[b]	0.86[b]	5.5	1.7[b]	—	—	33.9
	Cenchrus spp. Sandbur													
243	fresh, early bloom	2-04-177	20.0	14.0[a]	0.60[b]	0.49[b]	0.31[b]	0.18[b]	0.31[b]	2.1	1.2[b]	—	—	7.6
244			100.0	68.0[a]	3.00[b]	2.46[b]	1.52[b]	0.91[b]	1.55[b]	10.3	6.2[b]	—	—	37.8
	Chloris spp. Windmillgrass													
245	fresh	2-05-479	20.0	11.0[a]	0.49[b]	0.41[b]	0.24[b]	0.10[b]	0.25[b]	1.8	1.0[b]	—	—	5.8
246			100.0	56.0[a]	2.47[b]	2.03[b]	1.21[b]	0.51[b]	1.25[b]	9.1	5.1[b]	—	—	29.2
	Cichorium intybus. Chicory, Bluedaisy													
247	leaves, fresh	2-19-486	23.0	11.0	0.49[b]	0.40[b]	0.24[b]	0.04[b]	0.24[b]	5.0	4.2	—	—	2.5
248			100.0	48.0	2.12[b]	1.74[b]	1.03[b]	0.19[b]	1.06[b]	21.8	18.3	—	—	11.0
	Commelina spp. Dayflower													
249	fresh, late vegetative	2-30-040	17.0	8.0[b]	0.36[b]	0.29[b]	0.18[b]	0.04[b]	0.18[b]	2.0	1.3[b]	7.0	—	—
250			100.0	49.0[b]	2.16[b]	1.77[b]	1.05[b]	0.23[b]	1.08[b]	11.9	7.7[b]	41.0	—	—
	Crotalaria juncea. Crotalaria, Sunn													
251	fresh	2-13-955	20.0	12.0	0.55[b]	0.45[b]	0.27[b]	0.14[b]	0.28[b]	1.2	0.2	—	—	2.1
252			100.0	62.0	2.73[b]	2.24[b]	1.36[b]	0.72[b]	1.40[b]	6.0	0.8	—	—	10.5
	Croton spp. Croton													
253	browse, fresh	2-01-694	40.0	27.0[a]	1.18[b]	0.97[b]	0.60[b]	0.35[b]	0.61[b]	6.4	4.6[b]	—	—	9.2
254			100.0	67.0[a]	2.95[b]	2.42[b]	1.50[b]	0.88[b]	1.52[b]	16.0	11.5[b]	—	—	23.0
	Cucurbita foetidissima. Buffalogourd													
255	leaves, fresh	2-30-038	23.0	11.0[b]	0.49[b]	0.40[b]	0.24[b]	0.05[b]	0.24[b]	4.5	3.4[b]	4.0	—	—
256			100.0	49.0[b]	2.16[b]	1.77[b]	1.05[b]	0.23[b]	1.08[b]	19.9	15.2[b]	19.0	—	—
	Cyamopsis tetragonoloba. Guar													
257	fresh	2-02-333	26.0	16.0	0.72[b]	0.59[b]	0.36[b]	0.20[b]	0.37[b]	4.2	3.0	—	—	7.0
258			100.0	64.0	2.82[b]	2.31[b]	1.41[b]	0.78[b]	1.45[b]	16.4	11.8	—	—	27.4
	Cynodon dactylon. Bermudagrass													
259	fresh	2-00-712	34.0	20.0[a]	0.89[b]	0.73[b]	0.44[b]	0.22[b]	0.45[b]	4.1	2.6[b]	—	—	8.9
260			100.0	60.0[a]	2.65[b]	2.17[b]	1.31[b]	0.65[b]	1.35[b]	12.0	7.8[b]	—	—	26.4
261	fresh, late vegetative	2-10-131	38.0	23.0	1.03[b]	0.84[b]	0.51[b]	0.26[b]	0.53[b]	4.5	2.9[b]	24.0	—	—
262			100.0	61.0	2.69[b]	2.21[b]	1.33[b]	0.69[b]	1.38[b]	11.8	7.6[b]	64.0	—	—
	Dactylis glomerata. Orchardgrass													
263	fresh, early bloom	2-03-442	25.0	16.0[a]	0.72[b]	0.59[b]	0.36[b]	0.21[b]	0.37[b]	2.9	1.9[b]	13.0	8.0	7.9
264			100.0	66.0[a]	2.91[b]	2.39[b]	1.47[b]	0.85[b]	1.50[b]	11.8	7.6[b]	54.0	31.0	32.0
	Dactyloctenium aegyptium. Crowfootgrass, Durban													
265	fresh	2-21-442	17.0	11.0[a]	0.51[b]	0.42[b]	0.26[b]	0.16[b]	0.26[b]	2.3	1.6[b]	—	—	4.7
266			100.0	69.0[a]	3.04[b]	2.50[b]	1.55[b]	0.94[b]	1.57[b]	13.8	9.4[b]	—	—	28.0

TABLE 2 Composition of Goat Feeds—*Continued*

Entry Number	SCIENTIFIC NAME Feed Name	International Feed Number	Dry Matter (%)	Feed Energy TDN (%)	DE (Mcal/kg)	ME (Mcal/kg)	NE_m (Mcal/kg)	NE_g (Mcal/kg)	NE_l (Mcal/kg)	Protein Total (%)	Dig. (%)	Cell Walls (%)	Acid Detergent Fiber (%)	Crude Fiber (%)
	Dalbergia sissoo. Dalbergia, sissoo													
267	pods, fresh	2-30-174	17.0	4.0	0.16[b]	0.13[b]	0.13[b]	0.00[b]	0.07[b]	2.8	0.6	—	—	4.1
268			100.0	21.0	0.93[b]	0.76[b]	0.77[b]	0.00[b]	0.40[b]	16.2	3.7	—	—	24.2
	Dalea formosa. Dalea, feather													
269	browse, fresh, early	2-29-844	40.0	22.0[b]	0.95[b]	0.78[b]	0.47[b]	0.17[b]	0.48[b]	6.8	—	18.0	—	—
270	vegetative		100.0	54.0[b]	2.38[b]	1.95[b]	1.16[b]	0.43[b]	1.20[b]	17.1	—	46.0	—	—
	Dasylirion texanum. Sotol, Texas													
271	leaves, fresh	2-04-528	57.0	33.0[a]	1.46[b]	1.20[b]	0.72[b]	0.33[b]	0.74[b]	4.0	1.8[b]	—	—	22.2
272			100.0	58.0[a]	2.56[b]	2.10[b]	1.26[b]	0.58[b]	1.30[b]	7.1	3.2[b]	—	—	39.0
	Digera aneensis. Digera, aneensis													
273	fresh	2-30-149	13.0	9.0[a]	0.40[b]	0.33[b]	0.21[b]	0.13[b]	0.21[b]	3.7	3.0[b]	—	—	2.4
274			100.0	72.0[a]	3.17[b]	2.60[b]	1.64[b]	1.03[b]	1.64[b]	29.0	23.6[b]	—	—	18.9
	Diospyros texana. Persimmon, Texas													
275	leaves, fresh	2-29-858	48.0	26.0[b]	1.14[b]	0.94[b]	0.56[b]	0.21[b]	0.58[b]	5.7	3.6[b]	17.0	—	—
276			100.0	54.0[b]	2.38[b]	1.95[b]	1.16[b]	0.43[b]	1.20[b]	11.8	7.6[b]	36.0	—	—
	Echinochloa crusgalli. Barnyardgrass													
277	fresh	2-03-108	21.0	13.0[a]	0.55[b]	0.45[b]	0.28[b]	0.14[b]	0.28[b]	1.6	0.8[b]	—	—	5.9
278			100.0	61.0[a]	2.69[b]	2.21[b]	1.33[b]	0.69[b]	1.38[b]	8.0	4.0[b]	—	—	28.7
	Elymus canadensis. Wildrye, Canada													
279	fresh, late vegetative	2-29-848	48.0	22.0[b]	0.98[b]	0.80[b]	0.48[b]	0.05[b]	0.49[b]	4.1	2.2[b]	31.0	—	—
280			100.0	46.0[b]	2.03[b]	1.66[b]	0.99[b]	0.10[b]	1.01[b]	8.5	4.5[b]	64.0	—	—
	Elymus junceus. Wildrye, Russian													
281	fresh	2-05-469	33.0	22.0[a]	0.96[b]	0.79[b]	0.49[b]	0.28[b]	0.49[b]	4.7	3.2[b]	—	—	7.4
282			100.0	66.0[a]	2.91[b]	2.39[b]	1.47[b]	0.85[b]	1.50[b]	14.1	9.7[b]	—	—	22.4
	Engelmannia pinnatifida. Engelmann daisy													
283	fresh	2-29-879	30.0	13.0[b]	0.56[b]	0.46[b]	0.28[b]	0.00[b]	0.27[b]	2.8	1.5[b]	15.0	—	—
284			100.0	42.0[b]	1.85[b]	1.52[b]	0.92[b]	0.00[b]	0.91[b]	9.2	5.1[b]	49.0	—	—
	Eriochloa sericea. Cupgrass, Texas													
285	fresh, late vegetative	2-29-996	40.0	15.0[b]	0.67[b]	0.55[b]	0.34[b]	0.00[b]	0.32[b]	3.1	1.5[b]	27.0	—	—
286			100.0	38.0[b]	1.68[b]	1.37[b]	0.87[b]	0.00[b]	0.81[b]	7.8	3.8[b]	69.0	—	—
287	fresh, mature	2-30-059	83.0	25.0[b]	1.10[b]	0.90[b]	0.65[b]	0.00[b]	0.51[b]	4.1	1.0[b]	58.0	—	—
288			100.0	30.0[b]	1.32[b]	1.09[b]	0.78[b]	0.00[b]	0.62[b]	5.0	1.2[b]	70.0	—	—
	Eurotia lanata. Winterfat, common													
289	fresh, early vegetative	2-05-489	30.0	19.0[a]	0.85[b]	0.69[b]	0.42[b]	0.24[b]	0.43[b]	6.3	4.8[b]	—	—	7.8
290			100.0	64.0[a]	2.82[b]	2.31[b]	1.41[b]	0.78[b]	1.45[b]	21.0	16.2[b]	—	—	26.0

TABLE 2 Composition of Goat Feeds—*Continued*

Entry Number	SCIENTIFIC NAME Feed Name	International Feed Number	Dry Matter (%)	Feed Energy TDN (%)	DE (Mcal/kg)	ME (Mcal/kg)	NE_m (Mcal/kg)	NE_g (Mcal/kg)	NE_l (Mcal/kg)	Protein Total (%)	Dig. (%)	Cell Walls (%)	Acid Detergent Fiber (%)	Crude Fiber (%)
291	fresh, stem-cured	2-26-142	80.0	28.0[a]	1.24[b]	1.01[b]	0.66[b]	0.00[b]	0.59[b]	8.7	5.3[b]	58.0	35.0	—
292			100.0	35.0[a]	1.54[b]	1.27[b]	0.83[b]	0.00[b]	0.74[b]	10.8	6.7[b]	72.0	44.0	—
	Evax prolifera. Evax, prolifera													
293	fresh, early vegetative	2-30-068	38.0	18.0[b]	0.78[b]	0.64[b]	0.38[b]	0.04[b]	0.39[b]	4.6	3.0[b]	16.0	—	—
294			100.0	46.0[b]	2.03[b]	1.66[b]	0.99[b]	0.10[b]	1.01[b]	12.0	7.8[b]	42.0	—	—
	Ficus benghalensis. Banyantree													
295	leaves, fresh	2-27-208	32.0	16.0	0.72[b]	0.59[b]	0.35[b]	0.10[b]	0.36[b]	3.1	1.3	—	—	8.7
296			100.0	51.0	2.25[b]	1.84[b]	1.10[b]	0.32[b]	1.13[b]	9.6	4.1	—	—	27.0
	Ficus glomerata. Fig, cluster													
297	leaves, fresh	2-27-209	30.0	16.0	0.71[b]	0.59[b]	0.35[b]	0.13[b]	0.36[b]	3.4	2.0	—	—	3.7
298			100.0	54.0	2.38[b]	1.95[b]	1.16[b]	0.43[b]	1.20[b]	11.2	6.7[b]	—	—	12.3
	Ficus religiosa. Peepultree													
299	leaves, fresh	2-27-207	17.0	6.0	0.27[b]	0.22[b]	0.14[b]	0.00[b]	0.13[b]	2.4	1.3	—	—	3.8
300			100.0	36.0	1.59[b]	1.30[b]	0.84[b]	0.00[b]	0.76[b]	14.0	7.6	—	—	22.4
	Ficus roxburghii. Fig, Roxburgh													
301	fresh	2-30-177	33.0	19.0[a]	0.84[b]	0.69[b]	0.41[b]	0.18[b]	0.43[b]	4.5	2.1[a]	—	—	2.6
302			100.0	57.0[a]	2.51[b]	2.06[b]	1.23[b]	0.55[b]	1.28[b]	13.4	6.2[a]	—	—	7.7
	Ficus virens. Fig, spotted													
303	leaves, fresh	2-28-704	15.0	9.0	0.42[b]	0.34[b]	0.21[b]	0.11[b]	0.21[b]	1.7	1.0	—	—	4.2
304			100.0	62.0	2.73[b]	2.24[b]	1.36[b]	0.72[b]	1.40[b]	11.2	6.3	—	—	27.7
	Forestiera pubescens. Forestiera, downy													
305	browse, fresh, late	2-29-837	38.0	29.0[b]	1.26[b]	1.03[b]	0.66[b]	0.42[b]	0.65[b]	4.9	—	11.0	—	—
306	vegetative		100.0	75.0[b]	3.31[b]	2.71[b]	1.73[b]	1.11[b]	1.72[b]	12.9	—	30.0	—	—
	Grewia oppositifolia roxburghii. Grewia, oppositifolia, roxburghii													
307	browse, fresh	2-30-097	33.0	7.0[a]	0.29[b]	0.24[b]	0.25[b]	0.00[b]	0.12[b]	6.5	1.7[a]	—	—	6.5
308			100.0	20.0[a]	0.88[b]	0.72[b]	0.77[b]	0.00[b]	0.37[b]	19.7	5.1[a]	—	—	19.8
	Gutierrezia texana. Snakeweed, Texas													
309	fresh, late vegetative	2-04-245	30.0	14.0[a]	0.61[b]	0.50[b]	0.30[b]	0.03[b]	0.30[b]	4.4	3.1[b]	—	—	5.7
310			100.0	46.0[a]	2.03[b]	1.66[b]	0.99[b]	0.10[b]	1.01[b]	14.7	10.3[b]	—	—	19.0
	Gymnosperia spinosa. Gymnosperia, spinosa													
311	leaves, fresh	2-30-124	18.0	11.0	0.50[b]	0.41[b]	0.25[b]	0.14[b]	0.26[b]	1.7	1.1	—	—	2.8
312			100.0	63.0	2.78[b]	2.28[b]	1.39[b]	0.75[b]	1.42[b]	9.4	6.1	—	—	15.3
	Helianthus spp. Sunflower													
313	fresh, late vegetative	2-04-722	12.0	7.0[a]	0.33[b]	0.27[b]	0.16[b]	0.09[b]	0.17[b]	1.0	0.5[b]	—	—	3.9
314			100.0	64.0[a]	2.82[b]	2.31[b]	1.41[b]	0.78[b]	1.45[b]	8.3	4.3[b]	—	—	34.2

TABLE 2 Composition of Goat Feeds—*Continued*

Entry Number	SCIENTIFIC NAME Feed Name	International Feed Number	Dry Matter (%)	Feed Energy: TDN (%)	DE (Mcal/kg)	ME (Mcal/kg)	NE_m (Mcal/kg)	NE_g (Mcal/kg)	NE_l (Mcal/kg)	Protein: Total (%)	Dig. (%)	Cell Walls (%)	Acid Detergent Fiber (%)	Crude Fiber (%)
315	*Hilaria belangeri*. Curly mesquite browse, fresh, early vegetative	2-01-723	23.0	14.0[a]	0.60[b]	0.49[b]	0.30[b]	0.14[b]	0.31[b]	4.0	2.9[b]	—	—	5.9
316			100.0	59.0[a]	2.60[b]	2.13[b]	1.28[b]	0.62[b]	1.33[b]	17.2	12.6[b]	—	—	25.6
317	*Hordeum vulgare*. Barley fresh	2-00-511	21.0	13.0[a]	0.59[b]	0.49[b]	0.30[b]	0.16[b]	0.30[b]	4.3	3.3[b]	—	—	4.7
318			100.0	63.0[a]	2.78[b]	2.28[b]	1.39[b]	0.75[b]	1.42[b]	20.4	15.5[b]	—	—	22.1
319	*Juniperus ashei*. Juniper, ashe leaves, fresh	2-29-850	52.0	33.0[b]	1.45[b]	1.19[b]	0.73[b]	0.40[b]	0.75[b]	3.3	1.4[b]	18.0	—	—
320			100.0	64.0[b]	2.82[b]	2.31[b]	1.41[b]	0.78[b]	1.45[b]	6.5	2.6[b]	34.0	—	—
321	*Juniperus pinchotii*. Juniper, redberry leaves, fresh	2-29-851	53.0	32.0[b]	1.39[b]	1.14[b]	0.69[b]	0.34[b]	0.71[b]	3.9	1.9[b]	19.0	—	—
322			100.0	60.0[b]	2.65[b]	2.17[b]	1.31[b]	0.65[b]	1.35[b]	7.5	3.5[b]	35.0	—	—
323	*Kochia vestita*. Summer-cypress, gray fresh, mature	2-08-844	65.0	36.0[a]	1.61[b]	1.32[b]	0.79[b]	0.33[b]	0.81[b]	5.9	3.2[b]	—	—	15.6
324			100.0	56.0[a]	2.47[b]	2.03[b]	1.21[b]	0.51[b]	1.25[b]	9.0	5.0[b]	—	—	24.0
325	*Lactuca sativa*. Lettuce fresh	2-02-624	5.0	4.0[a]	0.17[b]	0.14[b]	0.09[b]	0.05[b]	0.09[b]	1.2	0.9[b]	—	—	0.6
326			100.0	70.0[a]	3.09[b]	2.53[b]	1.58[b]	0.97[b]	1.60[b]	22.0	17.0[b]	—	—	11.2
327	*Lasiurus sindicus*. Lasiurus, sindicus fresh	2-30-173	22.0	9.0	0.40[b]	0.33[b]	0.20[b]	0.00[b]	0.20[b]	1.3	0.7	—	—	8.4
328			100.0	41.0	1.81[b]	1.48[b]	0.91[b]	0.00[b]	0.89[b]	6.0	3.0	—	—	38.0
329	*Lepidium* spp. Pepperweed fresh, late vegetative	2-03-686	42.0	23.0[b]	1.01[b]	0.83[b]	0.49[b]	0.18[b]	0.51[b]	9.1	7.0[b]	19.0	—	5.0
330			100.0	54.0[b]	2.38[b]	1.95[b]	1.16[b]	0.43[b]	1.20[b]	21.4	16.5[b]	45.0	—	11.9
331	*Leptochloa dubia*. Sprangle-top, green fresh, late vegetative	2-30-031	52.0	24.0[b]	1.07[b]	0.88[b]	0.52[b]	0.08[b]	0.53[b]	2.9	0.9[b]	39.0	—	—
332			100.0	47.0[b]	2.07[b]	1.70[b]	1.01[b]	0.15[b]	1.03[b]	5.6	1.8[b]	75.0	—	—
333	*Leptoloma cognatum*. Witch-grass, fall fresh, late vegetative	2-30-039	50.0	23.0[b]	1.01[b]	0.83[b]	0.50[b]	0.05[b]	0.50[b]	2.7	0.8[b]	34.0	—	—
334			100.0	46.0[b]	2.03[b]	1.66[b]	0.99[b]	0.10[b]	1.01[b]	5.5	1.6[b]	68.0	—	—
335	*Lespedeza striata*. Lespedeza, common fresh, full bloom	2-02-541	35.0	18.0[a]	0.80[b]	0.65[b]	0.39[b]	0.12[b]	0.40[b]	5.1	3.5[b]	—	—	—
336			100.0	52.0[a]	2.29[b]	1.88[b]	1.12[b]	0.36[b]	1.15[b]	14.6	10.2[b]	—	—	—
337	*Lesquerella gordonii*. Bladder-pod, Gordon fresh, late vegetative	2-29-820	35.0	13.0[b]	0.55[b]	0.45[b]	0.29[b]	0.00[b]	0.26[b]	3.4	1.9[b]	11.0	—	—
338			100.0	36.0[b]	1.59[b]	1.30[b]	0.84[b]	0.00[b]	0.76[b]	9.7	5.6[b]	31.0	—	—

TABLE 2 Composition of Goat Feeds—*Continued*

Entry Number	SCIENTIFIC NAME Feed Name	International Feed Number	Dry Matter (%)	Feed Energy TDN (%)	DE (Mcal/kg)	ME (Mcal/kg)	NE_m (Mcal/kg)	NE_g (Mcal/kg)	NE_l (Mcal/kg)	Protein Total (%)	Protein Dig. (%)	Cell Walls (%)	Acid Detergent Fiber (%)	Crude Fiber (%)
	Leucaena retusa. Leadtree, littleleaf													
339	browse, fresh, early vegetative	2-02-491	27.0	18.0[a]	0.81[b]	0.66[b]	0.41[b]	0.25[b]	0.42[b]	4.8	3.5[b]	—	—	6.6
340			100.0	68.0[a]	3.00[b]	2.46[b]	1.52[b]	0.91[b]	1.55[b]	17.6	13.0[b]	—	—	24.4
	Lolium spp. Ryegrass													
341	fresh	2-04-062	24.0	15.0[a]	0.65[b]	0.53[b]	0.32[b]	0.17[b]	0.33[b]	2.5	1.5[b]	—	—	7.0
342			100.0	61.0[a]	2.69[b]	2.21[b]	1.33[b]	0.69[b]	1.38[b]	10.2	6.1[b]	—	—	29.2
	Lonicera albiflora. Honeysuckle, bush													
343	leaves, fresh	2-29-875	33.0	22.0[b]	0.99[b]	0.81[b]	0.51[b]	0.31[b]	0.51[b]	3.3	1.9[b]	6.0	—	—
344			100.0	69.0[b]	3.04[b]	2.50[b]	1.55[b]	0.94[b]	1.57[b]	10.0	5.9[b]	20.0	—	—
	Lotus spp. Trefoil													
345	fresh	2-01-757	15.0	10.0[a]	0.44[b]	0.36[b]	0.22[b]	0.12[b]	0.22[b]	2.7	2.0[b]	—	—	—
346			100.0	64.0[a]	2.82[b]	2.31[b]	1.41[b]	0.78[b]	1.45[b]	17.5	12.9[b]	—	—	—
	Lupinus texensis. Bluebonnet, Texas													
347	fresh, early vegetative	2-29-821	18.0	11.0[b]	0.46[b]	0.38[b]	0.23[b]	0.11[b]	0.24[b]	2.8	2.0[b]	4.0	—	—
348			100.0	59.0[b]	2.60[b]	2.13[b]	1.28[b]	0.62[b]	1.33[b]	16.1	11.5[b]	24.0	—	—
349	fresh, late vegetative	2-29-831	16.0	11.0[b]	0.47[b]	0.38[b]	0.24[b]	0.14[b]	0.24[b]	2.9	2.2[b]	4.0	—	—
350			100.0	66.0[b]	2.91[b]	2.39[b]	1.47[b]	0.85[b]	1.50[b]	18.0	13.4[b]	25.0	—	—
	Medicago sativa. Alfalfa													
351	fresh	2-00-196	24.0	13.0	0.59[b]	0.48[b]	0.29[b]	0.12[b]	0.30[b]	4.8	3.0	—	—	6.4
352			100.0	55.0	2.43[b]	1.99[b]	1.19[b]	0.47[b]	1.23[b]	19.7	12.1	—	—	26.2
353	fresh, late vegetative	2-00-181	21.0	13.0[a]	0.59[b]	0.49[b]	0.30[b]	0.16[b]	0.30[b]	4.3	3.2[b]	8.0	6.0	4.9
354			100.0	63.0[a]	2.78[b]	2.28[b]	1.39[b]	0.75[b]	1.42[b]	20.0	15.2[b]	38.0	29.0	23.0
355	fresh, full bloom	2-00-188	25.0	14.0[a]	0.61[b]	0.50[b]	0.30[b]	0.12[b]	0.31[b]	3.5	2.4[b]	13.0	9.0	7.7
356			100.0	55.0[a]	2.43[b]	1.99[b]	1.19[b]	0.47[b]	1.23[b]	14.0	9.6[b]	52.0	37.0	31.0
	Melia azedarach. Chinaberry													
357	leaves, fresh	2-27-210	21.0	14.0	0.61[b]	0.50[b]	0.31[b]	0.18[b]	0.32[b]	3.9	2.9	—	—	2.5
358			100.0	67.0	2.95[b]	2.42[b]	1.50[b]	0.88[b]	1.52[b]	18.7	13.9	—	—	11.9
	Mimosa spp. Mimosa													
359	browse, fresh, early vegetative	2-03-122	26.0	18.0[a]	0.79[b]	0.65[b]	0.40[b]	0.24[b]	0.41[b]	5.4	4.1[b]	—	—	5.5
360			100.0	69.0[a]	3.04[b]	2.50[b]	1.55[b]	0.94[b]	1.57[b]	20.6	15.8[b]	—	—	21.1
	Moringa oleifera. Horseradishtree													
361	leaves, fresh	2-30-172	18.0	11.0	0.48[b]	0.40[b]	0.24[b]	0.12[b]	0.25[b]	3.4	2.0	—	—	3.2
362			100.0	61.0	2.69[b]	2.21[b]	1.33[b]	0.69[b]	1.38[b]	19.0	11.0	—	—	17.9
	Morus australis. Mulberry, Japanese													
363	leaves, fresh	2-30-179	15.0	9.0[a]	0.40[b]	0.33[b]	0.20[b]	0.10[b]	0.20[b]	2.3	1.6[a]	—	—	2.3
364			100.0	60.0[a]	2.65[b]	2.17[b]	1.31[b]	0.65[b]	1.35[b]	15.0	10.7[a]	—	—	15.3
	Morus spp. Mulberry													
365	browse, fresh	2-03-150	40.0	29.0[a]	1.26[b]	1.03[b]	0.65[b]	0.40[b]	0.65[b]	7.3	5.4[b]	—	—	4.5
366			100.0	71.0[a]	3.13[b]	2.57[b]	1.61[b]	1.00[b]	1.62[b]	18.1	13.4[b]	—	—	11.2

TABLE 2 Composition of Goat Feeds—*Continued*

Entry Number	SCIENTIFIC NAME Feed Name	International Feed Number	Dry Matter (%)	Feed Energy TDN (%)	DE (Mcal/kg)	ME (Mcal/kg)	NE_m (Mcal/kg)	NE_g (Mcal/kg)	NE_l (Mcal/kg)	Protein Total (%)	Protein Dig. (%)	Cell Walls (%)	Acid Detergent Fiber (%)	Crude Fiber (%)
	Musa spp. Banana													
367	leaves, fresh	2-09-902	19.0	9.0	0.41[b]	0.34[b]	0.20[b]	0.05[b]	0.21[b]	2.9	1.3	—	—	4.4
368			100.0	50.0	2.21[b]	1.81[b]	1.07[b]	0.28[b]	1.11[b]	15.7	6.7	—	—	23.3
	Nolina texana. Nolina, Texas													
369	buds, fresh	2-29-855	32.0	20.0[b]	0.89[b]	0.73[b]	0.44[b]	0.24[b]	0.45[b]	6.2	4.7[b]	9.0	—	—
370			100.0	63.0[b]	2.78[b]	2.28[b]	1.39[b]	0.75[b]	1.42[b]	19.4	14.7[b]	28.0	—	—
371	leaves, fresh	2-29-852	57.0	27.0[b]	1.18[b]	0.97[b]	0.58[b]	0.08[b]	0.59[b]	3.2	1.0[b]	36.0	—	—
372			100.0	47.0[b]	2.07[b]	1.70[b]	1.01[b]	0.15[b]	1.03[b]	5.6	1.8[b]	63.0	—	—
	Onobrychis viciifolia. Sainfoin, common													
373	fresh	2-08-503	26.0	16.0[a]	0.72[b]	0.59[b]	0.36[b]	0.20[b]	0.37[b]	3.8	2.7[b]	—	—	6.2
374			100.0	64.0[a]	2.82[b]	2.31[b]	1.41[b]	0.78[b]	1.45[b]	14.8	10.4[b]	—	—	24.2
	Opuntia spp. Pricklypear													
375	fresh, mature	2-01-059	21.0	11.0[b]	0.48[b]	0.40[b]	0.24[b]	0.07[b]	0.24[b]	0.7	0.0[b]	4.0	—	2.9
376			100.0	51.0[b]	2.25[b]	1.84[b]	1.10[b]	0.32[b]	1.13[b]	3.1	0.0[b]	21.0	—	13.7
377	fruit, fresh, immature	4-30-020	26.0	9.0[b]	0.39[b]	0.32[b]	0.21[b]	0.00[b]	0.19[b]	1.8	0.9[b]	14.0	—	—
378			100.0	34.0[b]	1.50[b]	1.23[b]	0.82[b]	0.00[b]	0.71[b]	6.8	3.5[b]	52.0	—	—
	Panicum antidotale. Panicum, blue													
379	fresh	2-03-508	28.0	13.0	0.56[b]	0.46[b]	0.28[b]	0.02[b]	0.28[b]	3.7	1.4	—	—	9.1
380			100.0	45.0	1.98[b]	1.63[b]	0.98[b]	0.06[b]	0.98[b]	13.1	4.8	—	—	32.0
	Panicum hallii. Panicum, Halls													
381	fresh, late vegetative	2-29-998	49.0	23.0[b]	1.02[b]	0.84[b]	0.50[b]	0.07[b]	0.51[b]	3.6	—	32.0	—	—
382			100.0	47.0[b]	2.07[b]	1.70[b]	1.01[b]	0.15[b]	1.03[b]	7.4	—	66.0	—	—
	Panicum maximum. Guineagrass													
383	fresh, 15 to 28 days' growth	2-09-995	16.0	9.0[a]	0.41[b]	0.33[b]	0.20[b]	0.09[b]	0.21[b]	3.0	2.4	—	—	4.4
384			100.0	58.0[a]	2.56[b]	2.10[b]	1.26[b]	0.58[b]	1.30[b]	18.8	14.9	—	—	27.8
385	fresh, 29 to 42 days' growth	2-09-714	18.0	10.0[a]	0.45[b]	0.37[b]	0.22[b]	0.10[b]	0.23[b]	2.5	1.7	—	—	6.0
386			100.0	57.0[a]	2.51[b]	2.06[b]	1.23[b]	0.55[b]	1.28[b]	13.7	9.7	—	—	33.2
387	fresh, 43 to 56 days' growth	2-09-910	21.0	12.0[a]	0.51[b]	0.42[b]	0.25[b]	0.11[b]	0.26[b]	2.4	1.7	—	—	8.5
388			100.0	56.0[a]	2.47[b]	2.03[b]	1.21[b]	0.51[b]	1.25[b]	11.7	8.1	—	—	41.2
389	fresh, early bloom	2-02-339	24.0	13.0[a]	0.57[b]	0.46[b]	0.28[b]	0.10[b]	0.29[b]	1.4	0.4[b]	—	—	10.1
390			100.0	53.0[a]	2.34[b]	1.92[b]	1.14[b]	0.40[b]	1.18[b]	5.6	1.8[b]	—	—	41.8
391	fresh, full bloom	2-02-341	30.0	15.0[a]	0.68[b]	0.56[b]	0.33[b]	0.11[b]	0.34[b]	2.0	0.8[b]	—	—	10.2
392			100.0	52.0[a]	2.29[b]	1.88[b]	1.12[b]	0.36[b]	1.15[b]	6.6	2.7[b]	—	—	34.4
	Panicum obtusum. Vine mesquite													
393	fresh, late vegetative	2-30-009	47.0	24.0[b]	1.04[b]	0.85[b]	0.51[b]	0.13[b]	0.52[b]	3.3	1.5[b]	33.0	—	—
394			100.0	50.0[b]	2.21[b]	1.81[b]	1.07[b]	0.28[b]	1.11[b]	7.0	3.1[b]	70.0	—	—
	Panicum virgatum. Switchgrass													
395	fresh, early vegetative	2-04-795	35.0	21.0[a]	0.93[b]	0.76[b]	0.46[b]	0.24[b]	0.47[b]	3.7	2.3[b]	—	—	10.2
396			100.0	61.0[a]	2.69[b]	2.21[b]	1.33[b]	0.69[b]	1.38[b]	10.8	6.6[b]	—	—	29.7
397	fresh, mature	2-04-798	42.0	21.0[a]	0.91[b]	0.74[b]	0.44[b]	0.10[b]	0.45[b]	2.3	0.7[b]	—	—	18.2
398			100.0	49.0[a]	2.16[b]	1.77[b]	1.05[b]	0.23[b]	1.08[b]	5.5	1.7[b]	—	—	43.3

TABLE 2 Composition of Goat Feeds—*Continued*

Entry Number	SCIENTIFIC NAME Feed Name	International Feed Number	Dry Matter (%)	Feed Energy TDN (%)	DE (Mcal/kg)	ME (Mcal/kg)	NE_m (Mcal/kg)	NE_g (Mcal/kg)	NE_l (Mcal/kg)	Protein Total (%)	Protein Dig. (%)	Cell Walls (%)	Acid Detergent Fiber (%)	Crude Fiber (%)
	Paspalum dilatatum. Dallisgrass													
399	fresh, early vegetative	2-01-738	26.0	16.0[a]	0.71[b]	0.58[b]	0.35[b]	0.19[b]	0.36[b]	5.9	4.6[b]	—	—	7.7
400			100.0	63.0[a]	2.78[b]	2.28[b]	1.39[b]	0.75[b]	1.42[b]	23.2	18.2[b]	—	—	30.1
401	fresh, full bloom	2-01-739	30.0	17.0[a]	0.77[b]	0.63[b]	0.38[b]	0.18[b]	0.39[b]	2.1	1.0[b]	—	—	9.7
402			100.0	58.0[a]	2.56[b]	2.10[b]	1.26[b]	0.58[b]	1.30[b]	7.1	3.2[b]	—	—	32.2
	Pennisetum incomptum. Pennisetum, incomptum													
403	fresh	2-30-104	30.0	5.0[a]	0.24[b]	0.20[b]	0.24[b]	0.00[b]	0.10[b]	2.0	0.9[b]	—	—	10.1
404			100.0	18.0[a]	0.79[b]	0.65[b]	0.78[b]	0.00[b]	0.32[b]	6.7	2.8[b]	—	—	33.2
	Pennisetum purpureum. Napiergrass													
405	fresh	2-03-166	21.0	12.0[a]	0.51[b]	0.42[b]	0.25[b]	0.10[b]	0.26[b]	2.0	1.1[b]	—	—	7.2
406			100.0	55.0[a]	2.43[b]	1.99[b]	1.19[b]	0.47[b]	1.23[b]	9.2	5.2[b]	—	—	33.9
407	fresh, 15 to 28 days' growth	2-09-412	15.0	9.0	0.38[b]	0.32[b]	0.19[b]	0.09[b]	0.20[b]	1.7	1.3	—	—	4.7
408			100.0	60.0	2.65[b]	2.17[b]	1.31[b]	0.65[b]	1.35[b]	11.9	9.2	—	—	32.4
409	fresh, 29 to 42 days' growth	2-10-095	21.0	12.0	0.53[b]	0.44[b]	0.26[b]	0.12[b]	0.27[b]	2.1	1.6	—	—	8.9
410			100.0	58.0	2.56[b]	2.10[b]	1.26[b]	0.58[b]	1.30[b]	9.9	7.5	—	—	42.8
411	fresh, early bloom	2-03-159	15.0	9.0[a]	0.39[b]	0.32[b]	0.20[b]	0.10[b]	0.20[b]	1.6	1.0[b]	—	—	4.7
412			100.0	60.0[a]	2.65[b]	2.17[b]	1.31[b]	0.65[b]	1.35[b]	11.0	6.8[b]	—	—	31.5
	Pennisetum purpureum. Napiergrass													
413	fresh, post ripe	2-03-165	30.0	16.0[a]	0.71[b]	0.58[b]	0.35[b]	0.12[b]	0.36[b]	1.2	0.1[b]	—	—	12.0
414			100.0	53.0[a]	2.34[b]	1.92[b]	1.14[b]	0.40[b]	1.18[b]	4.1	0.4[b]	—	—	39.6
	Phalaris arundinacea. Canarygrass, reed													
415	fresh	2-01-113	27.0	17.0[a]	0.76[b]	0.62[b]	0.38[b]	0.22[b]	0.39[b]	3.1	2.0[b]	12.0	8.0	7.8
416			100.0	65.0[a]	2.87[b]	2.35[b]	1.44[b]	0.82[b]	1.47[b]	11.6	7.4[b]	46.0	28.0	29.5
	Phalaris tuberosa stenoptera. Hardinggrass													
417	fresh	2-02-354	27.0	16.0[a]	0.69[b]	0.57[b]	0.34[b]	0.16[b]	0.35[b]	4.6	3.4[b]	—	—	—
418			100.0	58.0[a]	2.56[b]	2.10[b]	1.26[b]	0.58[b]	1.30[b]	17.0	12.4[b]	—	—	—
	Phaseolus atropurpureus. Bean, purple													
419	leaves, fresh	2-30-175	15.0	8.0	0.35[b]	0.29[b]	0.17[b]	0.06[b]	0.18[b]	3.5	1.4	—	—	4.6
420			100.0	53.0	2.34[b]	1.92[b]	1.14[b]	0.40[b]	1.18[b]	23.0	9.2	—	—	30.4
	Pisum sativum sativum. Pea, garden													
421	fresh	2-27-582	16.0	11.0[a]	0.48[b]	0.39[b]	0.24[b]	0.15[b]	0.25[b]	2.4	1.7	—	—	—
422			100.0	68.0[a]	3.00[b]	2.46[b]	1.52[b]	0.91[b]	1.55[b]	14.8	10.4	—	—	—
	Pisum spp. Pea													
423	fresh, late vegetative	2-03-580	14.0	9.0[a]	0.40[b]	0.33[b]	0.20[b]	0.12[b]	0.21[b]	3.7	3.0[b]	—	—	3.0
424			100.0	67.0[a]	2.95[b]	2.42[b]	1.50[b]	0.88[b]	1.52[b]	27.4	22.1[b]	—	—	22.2
	Plantago rhodosperma. Plantain, redseed													
425	leaves, fresh	2-29-870	28.0	12.0[b]	0.54[b]	0.44[b]	0.27[b]	0.00[b]	0.26[b]	2.2	1.1[b]	6.0	—	—
426			100.0	43.0[b]	1.90[b]	1.56[b]	0.94[b]	0.00[b]	0.93[b]	7.7	3.7[b]	20.0	—	—

TABLE 2 Composition of Goat Feeds—*Continued*

Entry Number	SCIENTIFIC NAME Feed Name	International Feed Number	Dry Matter (%)	Feed Energy TDN (%)	DE (Mcal/kg)	ME (Mcal/kg)	NE_m (Mcal/kg)	NE_g (Mcal/kg)	NE_l (Mcal/kg)	Protein Total (%)	Dig. (%)	Cell Walls (%)	Acid Detergent Fiber (%)	Crude Fiber (%)
	Prosopis glandulosa. Mesquite, honey													
427	browse, fresh, late	2-29-999	48.0	22.0[b]	0.95[b]	0.78[b]	0.47[b]	0.03[b]	0.47[b]	7.8	5.6[b]	22.0	—	—
428	vegetative		100.0	45.0[b]	1.98[b]	1.63[b]	0.98[b]	0.06[b]	0.98[b]	16.2	11.7[b]	47.0	—	—
	Quercus sinuata breviloba. Oak, shin													
429	browse, fresh, early	2-29-826	32.0	23.0[b]	1.00[b]	0.82[b]	0.52[b]	0.32[b]	0.52[b]	5.5	4.0[b]	7.0	—	—
430	vegetative		100.0	72.0[b]	3.17[b]	2.60[b]	1.64[b]	1.03[b]	1.64[b]	17.4	12.8[b]	23.0	—	—
	Quercus virginiana. Oak, live													
431	browse, fresh, early	2-29-815	36.0	19.0[b]	0.85[b]	0.69[b]	0.41[b]	0.15[b]	0.43[b]	7.2	5.5[b]	12.0	—	—
432	vegetative		100.0	54.0[b]	2.38[b]	1.95[b]	1.16[b]	0.43[b]	1.20[b]	20.3	15.5[b]	34.0	—	—
433	leaves, fresh	2-29-859	50.0	23.0[b]	1.01[b]	0.83[b]	0.50[b]	0.05[b]	0.50[b]	5.1	3.0[b]	23.0	—	—
434			100.0	46.0[b]	2.03[b]	1.66[b]	0.99[b]	0.10[b]	1.01[b]	10.2	6.1[b]	45.0	—	—
	Ratibida columnifera. Prairieconeflower, upright													
435	fresh, early vegetative	2-29-846	18.0	9.0[b]	0.41[b]	0.34[b]	0.20[b]	0.06[b]	0.21[b]	3.6	2.7[b]	5.0	—	—
436			100.0	51.0[b]	2.25[b]	1.84[b]	1.10[b]	0.32[b]	1.13[b]	19.4	14.7[b]	28.0	—	—
	Rhus aromatica. Sumac, fragrant													
437	browse, fresh, early	2-29-827	40.0	30.0[b]	1.35[b]	1.10[b]	0.71[b]	0.46[b]	0.70[b]	5.4	3.7[b]	6.0	—	—
438	vegetative		100.0	77.0[b]	3.40[b]	2.78[b]	1.79[b]	1.17[b]	1.77[b]	13.7	9.3[b]	15.0	—	—
	Sarcobatus vermiculatus. Greasewood, black													
439	browse, fresh	2-20-083	35.0	19.0[a]	0.85[b]	0.70[b]	0.42[b]	0.17[b]	0.43[b]	5.4	3.8[b]	—	—	5.8
440			100.0	55.0[a]	2.43[b]	1.99[b]	1.19[b]	0.47[b]	1.23[b]	15.3	10.8[b]	—	—	16.5
	Schrankia roemeriana. Sensitivebrier, Roemer													
441	browse, fresh, late	2-29-860	32.0	23.0[b]	1.03[b]	0.84[b]	0.53[b]	0.34[b]	0.53[b]	10.3	8.5[b]	8.0	—	—
442	vegetative		100.0	73.0[b]	3.22[b]	2.64[b]	1.67[b]	1.06[b]	1.67[b]	32.3	26.7[b]	25.0	—	—
	Secale cereale. Rye													
443	fresh	2-04-018	24.0	17.0[a]	0.73[b]	0.60[b]	0.37[b]	0.23[b]	0.38[b]	3.8	2.7[b]	—	—	6.8
444			100.0	69.0[a]	3.04[b]	2.50[b]	1.55[b]	0.94[b]	1.57[b]	15.9	11.4[b]	—	—	28.5
	Sedum nuttallianum. Stonecrop, Nuttall													
445	fresh, early vegetative	2-29-836	10.0	4.0[b]	0.19[b]	0.16[b]	0.10[b]	0.00[b]	0.10[b]	0.6	0.2[b]	1.0	—	—
446			100.0	44.0[b]	1.94[b]	1.59[b]	0.96[b]	0.01[b]	0.96[b]	6.2	2.3[b]	11.0	—	—
	Sesbania cannabina. Sesbania, pea													
447	browse, fresh	2-27-161	19.0	12.0	0.51[b]	0.41[b]	0.25[b]	0.12[b]	0.26[b]	5.3	4.3	—	—	3.1
448			100.0	60.0	2.65[b]	2.17[b]	1.31[b]	0.65[b]	1.35[b]	27.5	22.3	—	—	16.0
	Setaria macrostachya. Bristlegrass, plains													
449	fresh, late vegetative	2-29-882	43.0	15.0[b]	0.68[b]	0.55[b]	0.36[b]	0.00[b]	0.32[b]	5.4	—	28.0	—	—
450			100.0	36.0[b]	1.59[b]	1.30[b]	0.84[b]	0.00[b]	0.76[b]	12.7	—	66.0	—	—
	Setaria spp. Millet													
451	fresh	2-20-742	22.0	14.0[a]	0.61[b]	0.50[b]	0.31[b]	0.17[b]	0.31[b]	6.0	4.8[b]	—	—	5.0
452			100.0	63.0[a]	2.78[b]	2.28[b]	1.39[b]	0.75[b]	1.42[b]	27.1	21.9[b]	—	—	22.9

TABLE 2 Composition of Goat Feeds—*Continued*

Entry Number	SCIENTIFIC NAME Feed Name	International Feed Number	Dry Matter (%)	Feed Energy TDN (%)	DE (Mcal/kg)	ME (Mcal/kg)	NE_m (Mcal/kg)	NE_g (Mcal/kg)	NE_l (Mcal/kg)	Protein Total (%)	Dig. (%)	Cell Walls (%)	Acid Detergent Fiber (%)	Crude Fiber (%)
	Sorghum bicolor. Sorghum													
453	fresh	2-04-317	35.0	23.0	1.02[b]	0.84[b]	0.51[b]	0.30[b]	0.52[b]	1.4	0.8	—	—	11.9
454			100.0	66.0	2.91[b]	2.39[b]	1.47[b]	0.85[b]	1.50[b]	4.0	2.4	—	—	34.0
455	fresh, mature	2-04-315	39.0	24.0[a]	1.05[b]	0.87[b]	0.52[b]	0.27[b]	0.54[b]	1.4	0.0[b]	—	—	10.7
456			100.0	61.0[a]	2.69[b]	2.21[b]	1.33[b]	0.69[b]	1.38[b]	3.6	0.0[b]	—	—	27.3
	Sorghum bicolor sudanense. Sorghum, Sudangrass													
457	fresh, early vegetative	2-04-484	18.0	12.0[a]	0.55[b]	0.45[b]	0.28[b]	0.17[b]	0.28[b]	3.0	2.2[b]	—	—	5.5
458			100.0	70.0[a]	3.09[b]	2.53[b]	1.58[b]	0.97[b]	1.60[b]	16.8	12.2[b]	—	—	30.9
459	fresh, midbloom	2-04-485	23.0	14.0[a]	0.63[b]	0.52[b]	0.32[b]	0.17[b]	0.32[b]	2.0	1.1[b]	—	—	8.2
460			100.0	63.0[a]	2.78[b]	2.28[b]	1.39[b]	0.75[b]	1.42[b]	8.8	4.7[b]	—	—	36.1
	Sorghum halepense. Sorghum, Johnsongrass													
461	fresh	2-04-412	24.0	15.0[a]	0.66[b]	0.54[b]	0.33[b]	0.18[b]	0.34[b]	2.9	1.9[b]	—	—	7.1
462			100.0	63.0[a]	2.78[b]	2.28[b]	1.39[b]	0.75[b]	1.42[b]	12.3	8.1[b]	—	—	29.9
	Stipa leucotricha. Needlegrass, Texas													
463	fresh, late vegetative	2-29-822	55.0	23.0[b]	0.99[b]	0.81[b]	0.50[b]	0.00[b]	0.48[b]	5.9	3.7[b]	36.0	—	—
464			100.0	41.0[b]	1.81[b]	1.48[b]	0.91[b]	0.00[b]	0.89[b]	10.8	6.7[b]	67.0	—	—
	Syzygium cumini. Java plum													
365	leaves, fresh	2-26-570	30.0	13.0	0.56[b]	0.46[b]	0.28[b]	0.00[b]	0.28[b]	2.3	0.0	—	—	4.7
466			100.0	43.0	1.90[b]	1.56[b]	0.94[b]	0.00[b]	0.93[b]	7.8	0.1	—	—	15.9
	Tillandsia usneoides. Spanish moss													
467	whole, fresh	2-04-648	41.0	24.0[a]	1.09[b]	0.89[b]	0.54[b]	0.27[b]	0.55[b]	2.6	1.0[b]	—	—	12.3
468			100.0	60.0[a]	2.65[b]	2.17[b]	1.31[b]	0.65[b]	1.35[b]	6.4	2.5[b]	—	—	29.9
	Tribulus terrestris. Puncturevine													
469	fresh	2-27-107	22.0	11.0[a]	0.48[b]	0.40[b]	0.24[b]	0.06[b]	0.24[b]	4.1	3.0[b]	—	—	8.1
470			100.0	50.0[a]	2.21[b]	1.81[b]	1.07[b]	0.28[b]	1.11[b]	18.5	13.8[b]	—	—	37.0
	Trifolium alexandrinum. Clover, Egyptian													
471	fresh	2-01-349	18.0	11.0	0.50[b]	0.41[b]	0.25[b]	0.14[b]	0.25[b]	3.0	2.4	—	—	4.3
472			100.0	64.0	2.82[b]	2.31[b]	1.41[b]	0.78[b]	1.45[b]	16.9	13.7	—	—	24.6
	Triodia albescens. Triodia, white													
473	fresh, late vegetative	2-30-011	43.0	23.0[b]	1.02[b]	0.84[b]	0.50[b]	0.19[b]	0.52[b]	3.3	—	30.0	—	—
474			100.0	54.0[b]	2.38[b]	1.95[b]	1.16[b]	0.43[b]	1.20[b]	7.7	—	70.0	—	—
	Vicia sativa. Vetch, common													
475	fresh	2-05-123	20.0	12.0[a]	0.53[b]	0.44[b]	0.26[b]	0.13[b]	0.27[b]	3.8	2.8[b]	—	—	5.5
476			100.0	59.0[a]	2.60[b]	2.13[b]	1.28[b]	0.62[b]	1.33[b]	18.6	13.9[b]	—	—	27.0
	Vigna sinensis. Cowpea, common													
477	fresh	2-01-655	15.0	10.0[a]	0.43[b]	0.35[b]	0.21[b]	0.12[b]	0.22[b]	2.4	1.8[b]	—	—	3.5
478			100.0	64.0[a]	2.82[b]	2.31[b]	1.41[b]	0.78[b]	1.45[b]	16.2	11.7[b]	—	—	23.3

TABLE 2 Composition of Goat Feeds—*Continued*

Entry Number	SCIENTIFIC NAME Feed Name	International Feed Number	Dry Matter (%)	Feed Energy TDN (%)	DE (Mcal/kg)	ME (Mcal/kg)	NE_m (Mcal/kg)	NE_g (Mcal/kg)	NE_l (Mcal/kg)	Protein Total (%)	Protein Dig. (%)	Cell Walls (%)	Acid Detergent Fiber (%)	Crude Fiber (%)
	Yucca spp. Yucca													
479	flowers, fresh	2-05-536	15.0	11.0[b]	0.47[b]	0.39[b]	0.24[b]	0.15[b]	0.25[b]	3.0	2.3[b]	2.0	—	2.0
480			100.0	70.0[b]	3.09[b]	2.53[b]	1.58[b]	0.97[b]	1.60[b]	19.7	14.9[b]	13.0	—	13.3
481	leaves, fresh, immature	2-30-050	41.0	18.0[b]	0.78[b]	0.64[b]	0.39[b]	0.00[b]	0.38[b]	3.0	1.4[b]	29.0	—	—
482			100.0	43.0[b]	1.90[b]	1.56[b]	0.94[b]	0.00[b]	0.93[b]	7.3	3.4[b]	69.0	—	—
	Zexmenia hispida. Zexmenia, orange													
483	leaves, fresh	2-30-051	39.0	17.0[b]	0.73[b]	0.60[b]	0.36[b]	0.00[b]	0.36[b]	4.5	2.9[b]	11.0	—	—
484			100.0	43.0[b]	1.90[b]	1.56[b]	0.94[b]	0.00[b]	0.93[b]	11.7	7.5[b]	27.0	—	—
	Ziziphus jujuba. Jujube, common													
485	browse, fresh	2-30-091	32.0	10.0	0.44[b]	0.36[b]	0.25[b]	0.00[b]	0.20[b]	2.7	1.0	—	—	9.6
486			100.0	31.0	1.37[b]	1.12[b]	0.79[b]	0.00[b]	0.64[b]	8.6	3.1	—	—	30.1
	SILAGES													
	Avena sativa. Oats													
487	silage	3-03-298	31.0	19.0[a]	0.84[b]	0.69[b]	0.42[b]	0.22[b]	0.43[b]	3.0	1.2[a]	—	—	9.7
488			100.0	62.0[a]	2.73[b]	2.24[b]	1.36[b]	0.72[b]	1.40[b]	9.6	3.8[a]	—	—	31.5
489	silage, milk stage	3-10-556	23.0	14.0[a]	0.63[b]	0.51[b]	0.31[b]	0.17[b]	0.32[b]	3.1	2.0[b]	—	—	6.9
490			100.0	63.0[a]	2.78[b]	2.28[b]	1.39[b]	0.75[b]	1.42[b]	14.0	8.9[b]	—	—	30.6
	Beta vulgaris altissima. Beet, sugar													
491	aerial part with crowns,	3-00-660	22.0	11.0[a]	0.50[b]	0.41[b]	0.25[b]	0.07[b]	0.25[b]	3.0	1.9[b]	—	—	3.1
492	silage		100.0	51.0[a]	2.25[b]	1.84[b]	1.10[b]	0.32[b]	1.13[b]	13.4	8.4[b]	—	—	13.7
	Digitaria decumbens. Pangolagrass													
493	silage, mature	3-03-494	32.0	18.0[a]	0.79[b]	0.65[b]	0.39[b]	0.16[b]	0.40[b]	2.9	1.4[b]	—	—	—
494			100.0	56.0[a]	2.47[b]	2.03[b]	1.21[b]	0.51[b]	1.25[b]	8.9	4.3[b]	—	—	—
	Glycine max. Soybean													
495	silage wilted	3-04-584	35.0	19.0[a]	0.84[b]	0.69[b]	0.41[b]	0.16[b]	0.43[b]	6.2	4.3[b]	—	—	9.5
496			100.0	55.0[a]	2.43[b]	1.99[b]	1.19[b]	0.47[b]	1.23[b]	18.0	12.5[b]	—	—	27.5
	Medicago sativa. Alfalfa													
497	silage, late vegetative	3-00-204	21.0	13.0[a]	0.59[b]	0.48[b]	0.29[b]	0.16[b]	0.30[b]	4.2	3.0[b]	12.0	10.0	7.3
498			100.0	63.0[a]	2.78[b]	2.28[b]	1.39[b]	0.75[b]	1.42[b]	20.0	14.4[b]	55.0	47.0	34.6
499	silage, full bloom	3-00-207	26.0	15.0[a]	0.68[b]	0.55[b]	0.33[b]	0.15[b]	0.34[b]	4.6	3.2[b]	—	10.0	9.2
500			100.0	58.0[a]	2.56[b]	2.10[b]	1.26[b]	0.58[b]	1.30[b]	17.5	12.1[b]	—	38.0	34.9
	Pennisetum purpureum. Napiergrass													
501	silage	3-03-170	26.0	11.0[a]	0.49[b]	0.40[b]	0.24[b]	0.00[b]	0.24[b]	1.2	0.1[b]	—	—	10.0
502			100.0	43.0[a]	1.90[b]	1.56[b]	0.94[b]	0.00[b]	0.93[b]	4.8	0.5[b]	—	—	39.1
	Pisum spp. Pea													
503	silage	3-03-590	24.0	13.0[a]	0.57[b]	0.47[b]	0.28[b]	0.10[b]	0.29[b]	6.2	4.7[b]	—	—	15.4
504			100.0	54.0[a]	2.38[b]	1.95[b]	1.16[b]	0.43[b]	1.20[b]	25.7	19.6[b]	—	—	64.0

TABLE 2 Composition of Goat Feeds—*Continued*

Entry Number	SCIENTIFIC NAME Feed Name	International Feed Number	Dry Matter (%)	Feed Energy TDN (%)	DE (Mcal/kg)	ME (Mcal/kg)	NE_m (Mcal/kg)	NE_g (Mcal/kg)	NE_l (Mcal/kg)	Protein Total (%)	Protein Dig. (%)	Cell Walls (%)	Acid Detergent Fiber (%)	Crude Fiber (%)
	Sorghum bicolor saccharatum. Sorghum, sorgo													
505	silage, dough stage	3-04-466	26.0	15.0[a]	0.66[b]	0.54[b]	0.32[b]	0.15[b]	0.34[b]	1.5	0.4[b]	—	—	7.9
506			100.0	58.0[a]	2.56[b]	2.10[b]	1.26[b]	0.58[b]	1.30[b]	5.8	1.5[b]	—	—	30.6
507	silage, mature	3-04-467	27.0	17.0[a]	0.74[b]	0.60[b]	0.37[b]	0.20[b]	0.38[b]	1.7	0.6[b]	—	—	7.3
508			100.0	63.0[a]	2.78[b]	2.28[b]	1.39[b]	0.75[b]	1.42[b]	6.6	2.2[b]	—	—	27.6
	Zea mays. Corn													
509	silage, milk stage	3-02-818	22.0	14.0[a]	0.63[b]	0.51[b]	0.31[b]	0.17[b]	0.32[b]	2.0	1.0[b]	—	8.0	6.8
510			100.0	64.0[a]	2.82[b]	2.31[b]	1.41[b]	0.78[b]	1.45[b]	8.9	4.3[b]	—	37.0	30.5
511	silage, dough stage	3-02-819	26.0	19.0[a]	0.82[b]	0.67[b]	0.42[b]	0.26[b]	0.42[b]	2.1	0.9[b]	—	8.0	6.5
512			100.0	70.0[a]	3.09[b]	2.53[b]	1.58[b]	0.97[b]	1.60[b]	7.8	3.3[b]	—	31.0	24.5
	ENERGY FEEDS													
	Animal													
513	tallow, bleached stabilized	4-07-880	99.0	193.0[a]	8.51[b]	6.98[b]	5.20[b]	3.98[b]	4.61[b]	—	—	—	—	—
514			100.0	195.0[a]	8.60[b]	7.05[b]	5.25[b]	4.02[b]	4.66[b]	—	—	—	—	—
	Avena sativa. Oats													
515	grain	4-03-309	89.0	68.0[a]	3.02[b]	2.47[b]	1.59[b]	1.04[b]	1.57[b]	11.8	8.4[b]	28.0	14.0	10.8
516			100.0	77.0[a]	3.40[b]	2.78[b]	1.79[b]	1.17[b]	1.77[b]	13.3	9.4[b]	32.0	16.0	12.1
	Beta vulgaris altissima. Beet, sugar													
517	molasses, more than 48% invert sugar, more than	4-00-668	78.0	61.0[a]	2.71[b]	2.22[b]	1.44[b]	0.95[b]	1.41[b]	6.6	3.9[b]	—	—	—
518	79.5 degrees brix		100.0	79.0[a]	3.48[b]	2.86[b]	1.85[b]	1.22[b]	1.82[b]	8.5	5.0[b]	—	—	—
519	pulp, dehydrated	4-00-669	91.0	67.0[a]	2.96[b]	2.43[b]	1.54[b]	0.98[b]	1.54[b]	8.8	5.5[b]	49.0	30.0	18.0
520			100.0	74.0[a]	3.26[b]	2.68[b]	1.70[b]	1.08[b]	1.69[b]	9.7	6.1[b]	54.0	33.0	19.8
521	roots, fresh	4-00-677	20.0	17.0[a]	0.76[b]	0.62[b]	0.42[b]	0.28[b]	0.40[b]	1.3	0.7[b]	—	—	1.1
522			100.0	88.0[a]	3.88[b]	3.18[b]	2.12[b]	1.45[b]	2.04[b]	6.8	3.5[b]	—	—	5.9
	Bos taurus. Cattle													
523	whey, dehydrated	4-01-182	93.0	76.0	3.33[b]	2.73[b]	1.78[b]	1.19[b]	1.74[b]	13.3	9.6[b]	—	—	0.2
524			100.0	81.0	3.57[b]	2.93[b]	1.91[b]	1.27[b]	1.87[b]	14.2	10.2[b]	—	—	0.2
	Brassica oleracea capitata. Cabbage													
525	fresh	4-01-046	9.0	8.0[a]	0.35[b]	0.28[b]	0.19[b]	0.13[b]	0.18[b]	1.9	1.5[b]	—	—	1.0
526			100.0	84.0[a]	3.70[b]	3.04[b]	2.00[b]	1.35[b]	1.94[b]	20.6	16.1[b]	—	—	10.5
	Brassica rapa rapa. Turnip													
527	roots, fresh	4-05-067	9.0	8.0[a]	0.35[b]	0.29[b]	0.19[b]	0.13[b]	0.19[b]	1.1	0.8[b]	4.0	3.0	1.1
528			100.0	86.0[a]	3.79[b]	3.11[b]	2.06[b]	1.40[b]	1.99[b]	11.8	8.1[b]	44.0	34.0	11.5
	Ceratonia siliqua. Carob bean													
529	pods with seeds	4-08-370	88.0	69.0[a]	3.06[b]	2.51[b]	1.62[b]	1.07[b]	1.59[b]	5.5	2.6[b]	—	—	8.7
530			100.0	79.0[a]	3.48[b]	2.86[b]	1.85[b]	1.22[b]	1.82[b]	6.3	3.0[b]	—	—	9.9
	Citrus spp. Citrus													
531	pomace, dehydrated or	4-08-652	90.0	66.0[a]	2.90[b]	2.38[b]	1.50[b]	0.95[b]	1.50[b]	6.2	3.2[b]	—	—	12.8
532	sun-cured (pulp)		100.0	73.0[a]	3.22[b]	2.64[b]	1.67[b]	1.06[b]	1.67[b]	6.9	3.5[b]	—	—	14.2
	Daucus spp. Carrot													
533	roots, fresh	4-01-145	12.0	10.0[a]	0.44[b]	0.36[b]	0.24[b]	0.16[b]	0.23[b]	1.2	0.7[b]	1.0	1.0	1.2
534			100.0	84.0[a]	3.70[b]	3.04[b]	2.00[b]	1.35[b]	1.94[b]	9.9	6.3[b]	9.0	8.0	9.7

TABLE 2 Composition of Goat Feeds—*Continued*

Entry Number	SCIENTIFIC NAME Feed Name	International Feed Number	Dry Matter (%)	Feed Energy TDN (%)	DE (Mcal/kg)	ME (Mcal/kg)	NE_m (Mcal/kg)	NE_g (Mcal/kg)	NE_l (Mcal/kg)	Protein Total (%)	Protein Dig. (%)	Cell Walls (%)	Acid Detergent Fiber (%)	Crude Fiber (%)
	Hordeum vulgare. Barley													
535	grain	4-00-549	88.0	74.0	3.27[b]	2.68[b]	1.76[b]	1.19[b]	1.71[b]	11.9	8.5[b]	—	—	5.0
536			100.0	84.0	3.70[b]	3.04[b]	2.00[b]	1.35[b]	1.94[b]	13.5	9.6[b]	—	—	5.7
	Ipomoea batatas. Sweetpotato													
537	tubers, fresh	4-04-788	33.0	27.0[a]	1.19[b]	0.98[b]	0.64[b]	0.42[b]	0.62[b]	1.7	0.6[b]	—	—	1.2
538			100.0	81.0[a]	3.57[b]	2.93[b]	1.91[b]	1.27[b]	1.87[b]	5.0	1.9[b]	—	—	3.6
	Laminariales (order)-fucales (order). Seaweed, kelp													
539	whole, sun-cured	4-04-190	89.0	29.0[a]	1.26[b]	1.03[b]	0.71[b]	0.00[b]	0.59[b]	8.8	5.6[b]	—	—	5.4
540			100.0	32.0[a]	1.41[b]	1.16[b]	0.80[b]	0.00[b]	0.66[b]	9.8	6.2[b]	—	—	6.1
	Malus spp. Apple													
541	pomace, wet	4-00-424	20.0	15.0[a]	0.67[b]	0.55[b]	0.35[b]	0.22[b]	0.35[b]	1.2	0.5[b]	—	—	3.5
542			100.0	74.0[a]	3.26[b]	2.68[b]	1.70[b]	1.08[b]	1.69[b]	5.6	2.4[b]	—	—	16.9
	Manihot esculenta. Cassava, common													
543	peelings, dehydrated	4-11-937	88.0	76.0[a]	3.35[b]	2.75[b]	1.82[b]	1.23[b]	1.76[b]	1.9	0.0[b]	—	—	3.0
544			100.0	86.0[a]	3.79[b]	3.11[b]	2.06[b]	1.40[b]	1.99[b]	2.2	0.0[b]	—	—	3.4
545	tubers, fresh	4-09-599	37.0	30.0[a]	1.31[b]	1.07[b]	0.70[b]	0.46[b]	0.68[b]	1.3	0.2[b]	—	—	1.7
546			100.0	80.0[a]	3.53[b]	2.89[b]	1.88[b]	1.24[b]	1.84[b]	3.6	0.5[b]	—	—	4.6
	Oryza sativa. Rice													
547	bran with germs	4-03-928	91.0	63.0[a]	2.80[b]	2.29[b]	1.43[b]	0.88[b]	1.45[b]	12.7	9.2[b]	30.0	16.0	11.6
548			100.0	70.0[a]	3.09[b]	2.53[b]	1.58[b]	0.97[b]	1.60[b]	14.1	10.1[b]	33.0	18.0	12.8
549	grain	4-03-939	89.0	71.0[a]	3.13[b]	2.57[b]	1.67[b]	1.11[b]	1.63[b]	8.0	4.9[b]	—	—	9.0
550			100.0	80.0[a]	3.53[b]	2.89[b]	1.88[b]	1.24[b]	1.84[b]	9.0	5.5[b]	—	—	10.2
	Pennisetum glaucum. Pearlmillet													
551	grain	4-03-118	90.0	84.0	3.71[b]	3.05[b]	2.07[b]	1.42[b]	1.96[b]	15.7	8.8	—	—	3.7
552			100.0	94.0	4.14[b]	3.40[b]	2.31[b]	1.59[b]	2.18[b]	17.5	9.8	—	—	4.2
	Quercus spp. Oak													
553	fruit, fresh	4-07-755	64.0	30.0[a]	1.32[b]	1.08[b]	0.65[b]	0.09[b]	0.66[b]	3.1	1.1[b]	—	—	8.8
554			100.0	47.0[a]	2.07[b]	1.70[b]	1.01[b]	0.15[b]	1.03[b]	4.8	1.7[b]	—	—	13.9
	Saccharum officinarum. Sugarcane													
555	molasses, more than 46% invert sugar, more than	4-04-696	75.0	54.0[a]	2.37[b]	1.94[b]	1.22[b]	0.77[b]	1.23[b]	4.4	1.9[b]	—	—	—
556	79.5 degrees brix		100.0	72.0[a]	3.17[b]	2.60[b]	1.64[b]	1.03[b]	1.64[b]	5.8	2.6[b]	—	—	—
	Secale cereale. Rye													
557	bran	4-04-022	91.0	56.0[a]	2.44[b]	2.00[b]	1.21[b]	0.62[b]	1.25[b]	15.9	12.0[b]	—	—	6.9
558			100.0	61.0[a]	2.69[b]	2.21[b]	1.33[b]	0.69[b]	1.38[b]	17.5	13.3[b]	—	—	7.6
559	grain	4-04-047	88.0	74.0[a]	3.24[b]	2.66[b]	1.75[b]	1.18[b]	1.70[b]	12.1	8.7[b]	—	—	2.2
560			100.0	84.0[a]	3.70[b]	3.04[b]	2.00[b]	1.35[b]	1.94[b]	13.8	9.9[b]	—	—	2.5
	Solanum tuberosum. Potato													
561	starch process residue,	4-03-778	86.0	70.0[a]	3.11[b]	2.55[b]	1.67[b]	1.11[b]	1.63[b]	3.7	1.1[b]	—	—	10.7
562	dehydrated		100.0	82.0[a]	3.62[b]	2.97[b]	1.94[b]	1.30[b]	1.89[b]	4.4	1.2[b]	—	—	12.4
563	tubers, fresh	4-03-787	23.0	19.0[a]	0.84[b]	0.69[b]	0.45[b]	0.30[b]	0.44[b]	2.2	1.4[b]	—	—	0.6
564			100.0	81.0[a]	3.57[b]	2.93[b]	1.91[b]	1.27[b]	1.87[b]	9.5	5.9[b]	—	—	2.4

TABLE 2 Composition of Goat Feeds—*Continued*

Entry Number	SCIENTIFIC NAME Feed Name	International Feed Number	Dry Matter (%)	Feed Energy TDN (%)	DE (Mcal/kg)	ME (Mcal/kg)	NE_m (Mcal/kg)	NE_g (Mcal/kg)	NE_l (Mcal/kg)	Protein Total (%)	Protein Dig. (%)	Cell Walls (%)	Acid Detergent Fiber (%)	Crude Fiber (%)
565	tubers, silage	4-03-768	25.0	20.0[a]	0.89[b]	0.73[b]	0.48[b]	0.32[b]	0.47[b]	1.9	1.0[b]	—	—	1.0
566			100.0	82.0[a]	3.62[b]	2.97[b]	1.94[b]	1.30[b]	1.89[b]	7.6	4.2[b]	—	—	4.0
	Sorghum bicolor subglabrescens. Sorghum, milo													
567	grain	4-04-444	89.0	78.0[a]	3.46[b]	2.84[b]	1.90[b]	1.29[b]	1.82[b]	10.2	6.8[b]	21.0	5.0	2.2
568			100.0	88.0[a]	3.88[b]	3.18[b]	2.12[b]	1.45[b]	2.04[b]	11.4	7.7[b]	23.0	5.0	2.5
	Triticum aestivum. Wheat													
569	bran	4-05-190	89.0	62.0	2.74[b]	2.25[b]	1.40[b]	0.86[b]	1.42[b]	15.2	11.9	46.0	14.0	10.0
570			100.0	70.0	3.09[b]	2.53[b]	1.58[b]	0.97[b]	1.60[b]	17.1	13.4	51.0	15.0	11.3
571	grain	4-05-211	89.0	78.0[a]	3.45[b]	2.83[b]	1.89[b]	1.28[b]	1.81[b]	14.2	10.6[b]	—	7.0	2.6
572			100.0	88.0[a]	3.88[b]	3.18[b]	2.12[b]	1.45[b]	2.04[b]	16.0	11.9[b]	—	8.0	2.9
	Zea mays. Corn													
573	ears, ground	4-02-849	87.0	71.0[a]	3.17[b]	2.60[b]	1.70[b]	1.14[b]	1.66[b]	7.8	4.8[b]	—	—	8.2
574			100.0	83.0[a]	3.66[b]	3.00[b]	1.97[b]	1.32[b]	1.91[b]	9.0	5.5[b]	—	—	9.4
575	grain	4-02-879	87.0	77.0	3.43[b]	2.81[b]	1.88[b]	1.28[b]	1.80[b]	9.2	7.5	—	—	2.1
576			100.0	89.0	3.92[b]	3.22[b]	2.15[b]	1.47[b]	2.06[b]	10.6	8.6	—	—	2.4
	Zea mays indentata. Corn, dent yellow													
577	grain	4-02-935	89.0	77.0[a]	3.40[b]	2.79[b]	1.85[b]	1.26[b]	1.78[b]	9.6	—	8.0	3.0	2.6
578			100.0	87.0[a]	3.84[b]	3.15[b]	2.09[b]	1.42[b]	2.01[b]	10.9	—	9.0	3.0	2.9
	PROTEIN SUPPLEMENTS													
	Arachis hypogaea. Peanut													
579	kernels	5-03-657	95.0	127.0[a]	5.60[b]	4.59[b]	3.32[b]	2.40[b]	3.00[b]	28.4	23.6[a]	—	—	2.8
580			100.0	134.0[a]	5.91[b]	4.85[b]	3.51[b]	2.54[b]	3.16[b]	29.9	24.9[a]	—	—	3.0
581	kernels, mechanical	5-03-648	94.0	80.0	3.55[b]	2.91[b]	1.93[b]	1.31[b]	1.86[b]	41.4	37.3	—	—	5.9
582	extracted caked		100.0	86.0	3.79[b]	3.11[b]	2.06[b]	1.40[b]	1.99[b]	44.3	39.8	—	—	6.3
583	kernels, meal solvent	5-03-650	92.0	70.0[a]	3.12[b]	2.56[b]	1.64[b]	1.07[b]	1.62[b]	48.1	—	—	—	9.9
584	extracted		100.0	77.0[a]	3.40[b]	2.78[b]	1.79[b]	1.17[b]	1.77[b]	52.3	—	—	—	10.8
	Blood. Meal spray dehy													
585	(blood flour)	5-00-381	93.0	85.0[a]	3.73[b]	3.06[b]	2.06[b]	1.41[b]	1.96[b]	82.8	78.9[a]	—	—	1.2
586			100.0	91.0[a]	4.01[b]	3.29[b]	2.22[b]	1.52[b]	2.11[b]	89.0	84.8[a]	—	—	1.3
	Bos Taurus. Cattle													
587	milk, fresh	5-01-168	12.0	16.0[a]	0.70[b]	0.58[b]	0.42[b]	0.30[b]	0.38[b]	3.3	3.1[a]	—	—	—
588			100.0	129.0[a]	5.69[b]	4.66[b]	3.37[b]	2.41[b]	3.04[b]	26.7	25.4[a]	—	—	—
	Brassica spp. Rape													
589	seeds, meal solvent	5-03-871	91.0	63.0[a]	2.77[b]	2.27[b]	1.41[b]	0.86[b]	1.43[b]	37.0	—	—	—	12.0
590	extracted		100.0	69.0[a]	3.04[b]	2.50[b]	1.55[b]	0.94[b]	1.57[b]	40.6	—	—	—	13.2
	Brevoortia tyrannus. Fish, menhaden													
591	meal mechanical extracted	5-02-009	92.0	67.0[a]	2.95[b]	2.42[b]	1.53[b]	0.97[b]	1.53[b]	61.1	49.4[a]	—	—	0.9
592			100.0	73.0[a]	3.22[b]	2.64[b]	1.67[b]	1.06[b]	1.67[b]	66.7	54.0[a]	—	—	1.0
	Carthamus tinctorius. Safflower													
593	seeds without hulls, meal	5-07-959	92.0	67.0[a]	2.95[b]	2.42[b]	1.53[b]	0.97[b]	1.53[b]	43.0	—	—	—	13.5
594	solvent extracted		100.0	73.0[a]	3.22[b]	2.64[b]	1.67[b]	1.06[b]	1.67[b]	46.9	—	—	—	14.7

TABLE 2 Composition of Goat Feeds—*Continued*

Entry Number	SCIENTIFIC NAME Feed Name	International Feed Number	Dry Matter (%)	Feed Energy TDN (%)	DE (Mcal/kg)	ME (Mcal/kg)	NE_m (Mcal/kg)	NE_g (Mcal/kg)	NE_l (Mcal/kg)	Protein Total (%)	Dig. (%)	Cell Walls (%)	Acid Detergent Fiber (%)	Crude Fiber (%)
	Cereals													
595	brewers grains, dehydrated	5-02-141	92.0	65.0[a]	2.84[b]	2.33[b]	1.46[b]	0.89[b]	1.47[b]	27.1	19.7[a]	42.0	22.0	13.2
596			100.0	70.0[a]	3.09[b]	2.53[b]	1.58[b]	0.97[b]	1.60[b]	29.4	21.4[a]	46.0	24.0	14.4
	Cicer arietinum. Chickpea													
597	seeds	5-01-218	91.0	69.0	3.04[b]	2.49[b]	1.59[b]	1.03[b]	1.58[b]	19.7	13.8	—	—	7.0
598			100.0	76.0	3.35[b]	2.75[b]	1.76[b]	1.14[b]	1.74[b]	21.7	15.2	—	—	7.7
	Cocos nucifera. Coconut													
599	meats oil residue, meal	5-01-573	91.0	69.0[a]	3.01[b]	2.47[b]	1.57[b]	1.01[b]	1.56[b]	21.3	—	—	—	14.0
600	solvent extracted		100.0	75.0[a]	3.31[b]	2.71[b]	1.73[b]	1.11[b]	1.72[b]	23.4	—	—	—	15.4
	Cyamopsis tetragonoloba. Guar													
601	seeds without endosperm,	5-05-687	88.0	73.0	3.24[b]	2.66[b]	1.75[b]	1.18[b]	1.70[b]	38.0	34.6	—	—	10.5
602	ground		100.0	84.0	3.70[b]	3.04[b]	2.00[b]	1.35[b]	1.94[b]	43.5	39.5	—	—	12.0
	Glycine max. Soybean													
603	seeds, meal solvent	5-04-604	90.0	79.0[a]	3.48[b]	2.85[b]	1.90[b]	1.30[b]	1.83[b]	44.8	41.5[a]	—	—	5.8
604	extracted		100.0	88.0[a]	3.88[b]	3.18[b]	2.12[b]	1.45[b]	2.04[b]	49.9	46.3[a]	—	—	6.5
	Gossypium spp. Cotton													
605	seeds, meal solvent ex-	5-01-621	91.0	69.0[a]	3.06[b]	2.51[b]	1.60[b]	1.04[b]	1.59[b]	41.2	—	—	16.0	12.1
606	tracted, 41% protein		100.0	76.0[a]	3.35[b]	2.75[b]	1.76[b]	1.14[b]	1.74[b]	45.2	—	—	17.0	13.3
	Helianthus annuus. Sunflower, common													
607	seeds without hulls, meal	5-04-739	93.0	60.0[a]	2.67[b]	2.19[b]	1.34[b]	0.76[b]	1.37[b]	46.3	—	—	—	11.4
608	solvent extracted		100.0	65.0[a]	2.87[b]	2.35[b]	1.44[b]	0.82[b]	1.47[b]	49.8	—	—	—	12.2
	Linum usitatissimum. Flax, common													
609	seeds, mechanical extracted	5-02-043	92.0	83.0	3.67[b]	3.01[b]	2.02[b]	1.38[b]	1.93[b]	32.3	27.5	—	—	9.2
610	caked		100.0	90.0	3.97[b]	3.25[b]	2.19[b]	1.49[b]	2.09[b]	35.0	29.8	—	—	10.0
611	seeds, meal solvent	5-02-048	90.0	70.0[a]	3.10[b]	2.54[b]	1.64[b]	1.08[b]	1.62[b]	34.6	29.6[a]	23.0	17.0	9.1
612	extracted		100.0	78.0[a]	3.44[b]	2.82[b]	1.82[b]	1.19[b]	1.79[b]	38.3	32.8[a]	25.0	19.0	10.1
	Lycopersicon esculentum. Tomato													
613	pomace, dehydrated	5-05-041	92.0	53.0[a]	2.35[b]	1.93[b]	1.16[b]	0.53[b]	1.19[b]	21.6	—	50.0	46.0	24.2
614			100.0	58.0[a]	2.56[b]	2.10[b]	1.26[b]	0.58[b]	1.30[b]	23.5	—	55.0	50.0	26.4
615	pomace, wet	5-05-042	13.0	8.0[a]	0.35[b]	0.29[b]	0.18[b]	0.09[b]	0.18[b]	2.8	—	—	—	3.5
616			100.0	62.0[a]	2.73[b]	2.24[b]	1.36[b]	0.72[b]	1.40[b]	21.7	—	—	—	27.3
	Meat, with bone, meal													
617	rendered	5-00-388	93.0	66.0[a]	2.92[b]	2.39[b]	1.50[b]	0.93[b]	1.51[b]	50.4	—	—	—	2.2
618			100.0	71.0[a]	3.13[b]	2.57[b]	1.61[b]	1.00[b]	1.62[b]	54.1	—	—	—	2.4
	Phaseolus vulgaris. Bean, navy													
619	seeds	5-00-623	89.0	75.0[a]	3.31[b]	2.72[b]	1.79[b]	1.21[b]	1.73[b]	22.6	20.2[a]	—	—	4.5
620			100.0	84.0[a]	3.70[b]	3.04[b]	2.00[b]	1.35[b]	1.94[b]	25.3	22.5[a]	—	—	5.0
	Poultry													
621	feathers, meal hydrolyzed	5-03-795	91.0	64.0[a]	2.82[b]	2.31[b]	1.44[b]	0.88[b]	1.46[b]	83.2	—	—	—	1.4
622			100.0	70.0[a]	3.09[b]	2.53[b]	1.58[b]	0.97[b]	1.60[b]	91.1	—	—	—	1.5

TABLE 2 Composition of Goat Feeds—*Continued*

Entry Number	SCIENTIFIC NAME Feed Name	International Feed Number	Dry Matter (%)	Feed Energy TDN (%)	DE (Mcal/kg)	ME (Mcal/kg)	NE_m (Mcal/kg)	NE_g (Mcal/kg)	NE_l (Mcal/kg)	Protein Total (%)	Protein Dig. (%)	Cell Walls (%)	Acid Detergent Fiber (%)	Crude Fiber (%)
	Secale cereale. Rye													
623	distillers grains, dehy-	5-04-023	92.0	56.0[a]	2.47[b]	2.03[b]	1.23[b]	0.63[b]	1.26[b]	21.6	12.9[a]	—	—	12.3
624	drated		100.0	61.0[a]	2.69[b]	2.21[b]	1.33[b]	0.69[b]	1.38[b]	23.5	14.1[a]	—	—	13.4
	Sesamum indicum. Sesame													
625	seeds, meal mechanical	5-04-220	93.0	71.0[a]	3.15[b]	2.58[b]	1.66[b]	1.08[b]	1.64[b]	45.5	—	16.0	16.0	5.7
626	extracted		100.0	77.0[a]	3.40[b]	2.78[b]	1.79[b]	1.17[b]	1.77[b]	49.1	—	17.0	17.0	6.1
	Sorghum bicolor. Sorghum													
627	distillers grains, dehy-	5-04-374	94.0	78.0[a]	3.43[b]	2.81[b]	1.85[b]	1.24[b]	1.79[b]	32.2	—	—	—	11.9
628	drated		100.0	83.0[a]	3.66[b]	3.00[b]	1.97[b]	1.32[b]	1.91[b]	34.4	—	—	—	12.7
	Triticum aestivum. Wheat													
629	germs, ground	5-05-218	88.0	83.0[a]	3.66[b]	3.00[b]	2.04[b]	1.40[b]	1.93[b]	24.8	—	—	—	3.1
630			100.0	94.0[a]	4.14[b]	3.40[b]	2.31[b]	1.59[b]	2.18[b]	28.1	—	—	—	3.5
	Urea													
631	45% nitrogen 281% pro-	5-05-070	99.0	—	0.00[b]	0.00[b]	0.00[b]	0.00[b]	0.00[b]	275.8	—	—	—	—
632	tein equivalent		100.0	—	0.00[b]	0.00[b]	0.00[b]	0.00[b]	0.00[b]	279.6	—	—	—	—

[a] Data derived with sheep.
[b] Calculated value (see p. 24).

TABLE 3 Composition of Goat Feeds: Dry Matter, Minerals, and Carotene Content; Data Expressed As-Fed and Dry (100% Dry Matter)

Entry Number	SCIENTIFIC NAME Feed Name	International Feed Number	Dry Matter (%)	Calcium (%)	Chlorine (%)	Magnesium (%)	Phosphorus (%)	Potassium (%)	Sodium (%)	Sulfur (%)	Cobalt (mg/kg)	Copper (mg/kg)	Iodine (mg/kg)	Iron (mg/kg)	Manganese (mg/kg)	Selenium (mg/kg)	Zinc (mg/kg)	Carotene (mg/kg)
	DRY FORAGES AND ROUGHAGES																	
	Agropyron desertorum. Wheatgrass, crested																	
001	hay, sun-cured	1-05-418	93.0	0.31	—	0.15	0.20	1.85	—	—	0.221	15.2	—	165.0	33.7	0.4	29.6	20.7
002			100.0	0.33	—	0.16	0.21	2.00	—	—	0.238	16.5	—	178.0	36.4	0.4	31.9	22.3
	Andropogon spp. Bluestem																	
003	hay, sun-cured	1-00-819	90.0	—	0.04	—	—	—	0.01	—	—	—	—	—	—	—	—	37.4
004			100.0	—	0.04	—	—	—	0.01	—	—	—	—	—	—	—	—	41.7
	Arachis hypogaea. Peanut																	
005	hay, sun-cured	1-03-619	91.0	1.12	—	0.45	0.14	1.25	—	0.21	0.072	—	—	—	—	—	—	31.6
006			100.0	1.23	—	0.49	0.15	1.38	—	0.23	0.079	—	—	—	—	—	—	34.8
007	hay with fruit, sun-cured	1-03-620	92.0	1.13	—	0.33	0,14	0.85	—	—	—	—	—	—	—	—	—	—
008			100.0	1.24	—	0.36	0.15	0.92	—	—	—	—	—	—	—	—	—	—
009	pods	1-08-028	91.0	0.24	—	0.15	0.06	0.87	0.12	0.09	0.108	16.3	—	285.0	62.6	—	22.0	0.8
010			100.0	0.26	—	0.17	0.07	0.95	0.13	0.10	0.119	17.8	—	312.0	68.7	—	24.1	0.9
	Avena sativa. Oats																	
011	hay, sun-cured	1-03-280	91.0	0.22	0.48	0.24	0.20	1.38	0.17	0.22	0.066	14.1	—	142.0	58.6	0.2	35.8	25.3
012			100.0	0.24	0.52	0.26	0.22	1.51	0.18	0.25	0.073	15.5	—	155.0	64.1	0.2	39.2	27.7
013	straw	1-03-283	92.0	0.22	0.71	0.17	0.06	2.37	0.39	0.21	—	9.5	—	161.0	33.7	—	5.7	3.5
014			100.0	0.24	0.78	0.18	0.06	2.57	0.42	0.23	—	10.3	—	175.0	36.6	—	6.2	3.8
	Brachiaria mutica. Paragrass																	
015	hay, sun-cured	1-03-517	90.0	0.32	—	—	0.29	1.43	—	—	—	—	—	—	—	—	—	—
016			100.0	0.35	—	—	0.32	1.60	—	—	—	—	—	—	—	—	—	—
	Bromus inermis. Brome, smooth																	
017	hay, sun-cured, early	1-00-941	91.0	0.24	—	0.16	0.20	2.11	0.01	—	0.255	14.1	—	112.0	54.5	0.5	26.2	—
018	bloom		100.0	0.27	—	0.18	0.22	2.31	0.01	—	0.280	15.4	—	123.0	59.7	0.5	28.8	—
019	hay, sun-cured, mature	1-00-944	93.0	0.29	0.12	0.12	0.20	1.85	0.01	—	0.133	6.3	—	93.0	98.0	—	22.2	—
020			100.0	0.32	0.13	0.13	0.22	2.00	0.01	—	0.143	6.8	—	100.0	105.8	—	24.0	—
	Carthamus tinctorius. Safflower																	
021	hay, sun-cured	1-30-127	85.0	1.19	—	0.37	0.18	—	—	—	—	—	—	—	—	—	—	—
022			100.0	1.40	—	0.43	0.21	—	—	—	—	—	—	—	—	—	—	—
	Chloris gayana. Rhodesgrass																	
023	hay, sun-cured, early	1-03-910	88.0	0.52	—	0.31	0.36	2.00	—	—	—	—	—	—	—	—	—	—
024	vegetative		100.0	0.59	—	0.35	0.41	2.27	—	—	—	—	—	—	—	—	—	—
	Cynodon dactylon. Bermudagrass																	
025	hay, sun-cured	1-00-703	91.0	0.43	—	0.16	0.16	1.40	0.07	0.19	0.111	—	0.11	265.0	—	—	—	52.7
026			100.0	0.47	—	0.17	0.17	1.53	0.08	0.21	0.121	—	0.12	290.0	—	—	—	57.6
	Cynodon dactylon. Bermudagrass, coastal																	
027	hay, sun-cured, late	1-20-900	91.0	—	—	—	—	—	—	—	—	—	—	—	—	—	—	116.0
028	vegetative		100.0	—	—	—	—	—	—	—	—	—	—	—	—	—	—	127.5
	Dactylis glomerata. Orchardgrass																	
029	hay, sun-cured, cut 1	1-03-433	92.0	0.31	—	0.12	0.31	3.31	0.07	—	0.427	11.5	—	112.0	95.7	—	24.9	—
030			100.0	0.34	—	0.13	0.34	3.58	0.08	—	0.463	12.5	—	121.0	103.7	—	27.0	—
031	hay, sun-cured, cut 2	1-03-434	89.0	0.41	—	0.17	0.42	3.28	0.04	—	0.455	11.8	—	68.0	118.7	—	22.3	—
032			100.0	0.46	—	0.19	0.47	3.69	0.04	—	0.511	13.3	—	77.0	113.4	—	25.1	—

TABLE 3 Composition of Goat Feeds—*Continued*

Entry Number	SCIENTIFIC NAME Feed Name	International Feed Number	Dry Matter (%)	Calcium (%)	Chlorine (%)	Magnesium (%)	Phosphorus (%)	Potassium (%)	Sodium (%)	Sulfur (%)	Cobalt (mg/kg)	Copper (mg/kg)	Iodine (mg/kg)	Iron (mg/kg)	Manganese (mg/kg)	Selenium (mg/kg)	Zinc (mg/kg)	Carotene (mg/kg)
	Echinochloa crusgalli. Barnyardgrass																	
033	hay, sun-cured	1-03-105	88.0	0.38	—	—	0.21	2.13	—	—	—	—	—	—	—	—	—	—
034			100.0	0.43	—	—	0.24	2.42	—	—	—	—	—	—	—	—	—	—
	Eragrostis spp. Lovegrass																	
035	hay, sun-cured, full	1-02-644	92.0	0.28	—	0.06	0.11	—	—	—	—	—	—	—	—	—	—	—
036	bloom		100.0	0.30	—	0.06	0.12	—	—	—	—	—	—	—	—	—	—	—
	Festuca arundinacea. Fescue, Kentucky 31																	
037	hay, sun-cured, early	1-09-186	91.0	0.35	—	0.11	0.24	1.78	0.02	—	—	15.5	—	221.0	73.7	—	31.9	—
038	bloom		100.0	0.38	—	0.12	0.26	1.96	0.02	—	—	17.0	—	243.0	81.0	—	35.0	—
	Glycine max. Soybean																	
039	hay, sun-cured, mid-	1-04-538	94.0	1.18	—	0.74	0.25	0.91	0.11	0.24	—	—	—	281.0	—	—	—	31.2
040	bloom		100.0	1.26	—	0.79	0.27	0.97	0.12	0.26	—	—	—	300.0	—	—	—	33.3
041	hay, sun-cured, mature	1-04-543	90.0	0.94	—	—	0.25	0.73	—	—	—	—	—	—	—	—	—	6.7
042			100.0	1.04	—	—	0.28	0.81	—	—	—	—	—	—	—	—	—	7.4
	Gossypium spp. Cotton																	
043	gin by-product	1-08-413	90.0	0.83	—	—	0.18	—	—	—	—	—	—	—	—	—	—	—
044			100.0	0.92	—	—	0.20	—	—	—	—	—	—	—	—	—	—	—
045	hulls	1-01-599	91.0	0.13	0.02	0.13	0.09	0.79	0.02	0.08	0.018	12.0	—	119.0	108.0	—	19.9	—
046			100.0	0.15	0.02	0.14	0.09	0.87	0.02	0.09	0.019	13.3	—	131.0	119.2	—	22.0	—
	Hordeum vulgare. Barley																	
047	straw	1-00-498	91.0	0.27	0.61	0.21	0.07	2.16	0.13	0.16	0.060	4.9	—	183.0	15.1	—	6.8	2.1
048			100.0	0.30	0.67	0.23	0.07	2.37	0.14	0.17	0.066	5.4	—	201.0	16.6	—	7.4	2.3
	Lespedeza striata. Lespedeza, common																	
049	hay, sun-cured, full bloom	1-20-887	89.0	1.02	—	0.20	0.19	0.93	—	—	—	—	—	268.0	101.9	—	36.3	—
050			100.0	1.14	—	0.23	0.21	1.04	—	—	—	—	—	300.0	114.1	—	40.7	—
	Lolium multiflorum. Ryegrass, Italian																	
051	hay, sun-cured, late	1-04-065	86.0	0.53	—	—	0.29	1.34	—	—	—	—	—	274.0	—	—	—	248.2
052	vegetative		100.0	0.62	—	—	0.34	1.56	—	—	—	—	—	320.0	—	—	—	290.0
	Medicago sativa. Alfalfa																	
053	meal, 17% protein	1-00-023	92.0	1.40	0.47	0.29	0.23	2.39	0.10	0.22	0.302	9.7	0.15	405.0	31.0	0.3	19.3	120.2
054			100.0	1.52	0.52	0.32	0.25	2.60	0.11	0.24	0.329	10.6	0.16	441.0	33.8	0.4	21.1	131.1
055	hay, sun-cured, early	1-00-050	90.0	1.62	0.31	0.23	0.32	1.99	0.20	0.57	0.090	9.5	—	228.0	40.9	—	21.6	180.9
056	vegetative		100.0	1.80	0.34	0.26	0.35	2.21	0.22	0.63	0.100	10.6	—	253.0	45.4	—	24.0	201.0
057	hay, sun-cured, late	1-00-054	90.0	1.38	0.31	0.22	0.26	2.29	0.13	0.28	0.081	7.9	—	204.0	30.5	—	24.6	181.2
058	vegetative		100.0	1.54	0.34	0.24	0.29	2.56	0.15	0.31	0.090	8.8	—	227.0	34.0	—	27.4	202.0
059	hay, sun-cured, early	1-00-059	90.0	1.27	0.34	0.29	0.20	2.27	0.13	0.25	0.145	9.9	—	173.0	27.5	0.5	22.2	126.1
060	bloom		100.0	1.41	0.38	0.33	0.22	2.52	0.14	0.28	0.161	10.9	—	192.0	30.5	0.5	24.6	140.1
061	hay, sun-cured, full	1-00-068	90.0	1.13	—	0.28	0.20	1.38	0.10	0.25	0.293	12.6	—	135.0	33.6	—	22.3	58.5
062	bloom		100.0	1.25	—	0.31	0.22	1.53	0.11	0.27	0.325	14.0	—	150.0	37.3	—	24.8	65.0
063	hay, sun-cured, mature	1-00-071	91.0	1.03	—	0.24	0.17	1.62	0.08	0.23	0.082	12.8	—	139.0	40.2	—	21.5	10.6
064			100.0	1.13	—	0.27	0.18	1.78	0.08	0.25	0.090	14.0	—	153.0	44.1	—	23.6	11.6
	Oryza sativa. Rice																	
065	straw	1-03-925	91.0	0.19	—	0.10	0.07	1.19	0.28	—	—	—	—	—	313.0	—	—	—
066			100.0	0.21	—	0.11	0.08	1.32	0.31	—	—	—	—	—	345.8	—	—	—
	Panicum maximum. Guineagrass																	
067	dehydrated pelleted, 15	1-30-084	92.0	0.44	—	0.23	0.19	—	—	—	—	—	—	—	—	—	—	—
068	to 28 days' growth		100.0	0.48	—	0.25	0.21	—	—	—	—	—	—	—	—	—	—	—

TABLE 3 Composition of Goat Feeds—*Continued*

Entry Number	SCIENTIFIC NAME Feed Name	International Feed Number	Dry Matter (%)	Calcium (%)	Chlorine (%)	Magnesium (%)	Phosphorus (%)	Potassium (%)	Sodium (%)	Sulfur (%)	Cobalt (mg/kg)	Copper (mg/kg)	Iodine (mg/kg)	Iron (mg/kg)	Manganese (mg/kg)	Selenium (mg/kg)	Zinc (mg/kg)	Carotene (mg/kg)
	Paspalum dilatatum. Dallisgrass																	
069	hay, sun-cured, mid-	1-01-734	91.0	0.39	—	—	0.15	—	—	—	—	—	—	—	—	—	—	—
070	bloom		100.0	0.43	—	—	0.17	—	—	—	—	—	—	—	—	—	—	—
	Paspalum notatum. Bahiagrass																	
071	hay, sun-cured	1-00-462	91.0	0.46	—	0.17	0.20	—	—	—	—	—	—	55.0	—	—	—	—
072			100.0	0.50	—	0.19	0.22	—	—	—	—	—	—	60.0	—	—	—	—
	Phalaris arundinacea. Canarygrass, reed																	
073	hay, sun-cured	1-01-104	91.0	0.35	—	0.27	0.23	2.51	0.13	—	0.020	10.8	—	137.0	107.7	—	—	23.4
074			100.0	0.38	—	0.29	0.25	2.76	0.14	—	0.022	11.9	—	150.0	118.3	—	—	25.7
	Phleum pratense. Timothy																	
075	hay, sun-cured, late	1-04-881	89.0	0.59	—	0.13	0.30	1.50	0.16	—	—	23.0	—	179.0	79.5	—	59.8	111.6
076	vegetative		100.0	0.66	—	0.14	0.34	1.68	0.18	—	—	25.8	—	200.0	89.0	—	67.0	125.0
077	hay, sun-cured, full	1-04-884	89.0	0.38	0.55	0.12	0.18	1.45	0.16	—	—	4.4	—	139.0	—	—	47.9	—
078	bloom		100.0	0.43	0.62	0.14	0.20	1.64	0.18	—	—	5.0	—	157.0	—	—	54.0	—
	Poa pratensis. Bluegrass, Kentucky																	
079	hay, sun-cured	1-00-776	89.0	0.29	0.47	0.14	0.22	1.51	0.12	0.14	—	8.8	—	261.0	61.9	—	—	—
080			100.0	0.33	0.53	0.16	0.25	1.69	0.13	0.16	—	9.9	—	293.0	69.5	—	—	—
	Prosopis spicigera. Mesquite, spicigera																	
081	hay, sun-cured	1-30-160	86.0	1.54	—	—	0.14	—	—	—	—	—	—	—	—	—	—	—
082			100.0	1.80	—	—	0.16	—	—	—	—	—	—	—	—	—	—	—
	Pueraria spp. Kudzu																	
083	hay, sun-cured	1-02-478	91.0	2.15	—	0.73	0.32	—	—	—	—	—	—	—	—	—	—	40.1
084			100.0	2.35	—	0.80	0.35	—	—	—	—	—	—	—	—	—	—	43.9
	Saccharum officinarum. Sugarcane																	
085	bagasse, dehydrated	1-04-686	91.0	0.82	—	0.09	0.27	0.46	0.18	0.09	—	—	—	91.0	—	—	—	—
086			100.0	0.90	—	0.10	0.29	0.50	0.20	0.10	—	—	—	100.0	—	—	—	—
	Secale cereale. Rye																	
087	hay, sun-cured	1-04-004	93.0	0.31	—	0.12	0.18	1.26	0.03	—	—	—	—	—	—	—	31.7	6.1
088			100.0	0.33	—	0.13	0.19	1.35	0.03	—	—	—	—	—	—	—	33.9	6.5
089	straw	1-04-007	90.0	0.22	0.21	0.07	0.08	0.87	0.12	0.10	—	3.6	—	—	6.0	—	—	—
090			100.0	0.24	0.24	0.08	0.09	0.97	0.13	0.11	—	4.0	—	—	6.6	—	—	—
	Sorghum bicolor. Sorghum																	
091	hay, sun-cured, full	1-04-371	90.0	0.56	—	0.27	0.17	1.11	0.02	—	—	—	—	—	—	—	—	—
092	bloom		100.0	0.62	—	0.30	0.19	1.24	0.02	—	—	—	—	—	—	—	—	—
093	aerial part, sun-cured,	1-04-301	90.0	0.56	—	0.27	0.17	1.12	0.02	—	—	—	—	181.0	—	—	—	—
094	mature		100.0	0.62	—	0.30	0.19	1.24	0.02	—	—	—	—	200.0	—	—	—	—
	Sorghum bicolor sudanense. Sorghum, Sudangrass																	
095	hay, sun-cured, early	1-04-473	90.0	0.69	—	0.27	0.32	1.92	0.01	—	—	31.4	—	565.0	71.0	—	44.9	—
096	vegetative		100.0	0.77	—	0.31	0.36	2.14	0.01	—	—	35.0	—	629.0	79.0	—	50.0	—
097	hay, sun-cured, late	1-04-474	88.0	0.38	—	0.12	0.26	2.83	0.01	—	—	44.8	—	367.0	43.1	—	53.6	—
098	vegetative		100.0	0.43	—	0.14	0.30	3.22	0.01	—	—	51.0	—	417.0	49.0	—	61.0	—
	Sorghum halepense. Sorghum, Johnsongrass																	
099	hay, sun-cured	1-04-407	89.0	0.75	—	0.31	0.25	1.21	0.01	0.09	—	—	—	527.0	—	—	—	34.8
100			100.0	0.84	—	0.35	0.28	1.35	0.01	0.10	—	—	—	590.0	—	—	—	38.9

TABLE 3 Composition of Goat Feeds—*Continued*

Entry Number	SCIENTIFIC NAME Feed Name	International Feed Number	Dry Matter (%)	Calcium (%)	Chlorine (%)	Magnesium (%)	Phosphorus (%)	Potassium (%)	Sodium (%)	Sulfur (%)	Cobalt (mg/kg)	Copper (mg/kg)	Iodine (mg/kg)	Iron (mg/kg)	Manganese (mg/kg)	Selenium (mg/kg)	Zinc (mg/kg)	Carotene (mg/kg)
	Trifolium alexandrinum. Clover, Egyptian																	
101	hay, sun-cured	1-01-340	88.0	2.25	—	0.35	0.26	2.00	—	—	—	—	—	—	—	—	—	—
102			100.0	2.54	—	0.40	0.30	2.26	—	—	—	—	—	—	—	—	—	—
	Trifolium pratense. Clover, red																	
103	hay, sun-cured	1-01-415	89.0	1.35	0.28	0.38	0.22	1.44	0.16	0.15	0.138	9.7	0.22	163.0	64.8	—	15.2	17.6
104			100.0	1.53	0.32	0.43	0.25	1.62	0.19	0.17	0.156	10.9	0.25	184.0	73.3	—	17.2	19.8
	Triticum aestivum. Wheat																	
105	straw	1-05-175	89.0	0.16	0.28	0.11	0.04	1.26	0.13	0.17	0.040	3.2	—	140.0	36.3	—	5.7	2.0
106			100.0	0.18	0.32	0.12	0.05	1.42	0.14	0.19	0.045	3.6	—	157.0	40.9	—	6.5	2.2
	Vicia spp. Vetch																	
107	hay, sun-cured	1-05-106	89.0	1.05	—	0.22	0.29	2.07	0.46	0.13	0.316	8.8	0.44	374.0	65.0	—	—	410.5
108			100.0	1.18	—	0.25	0.32	2.32	0.52	0.15	0.355	9.9	0.49	420.0	72.9	—	—	460.5
	Vigna sinensis. Cowpea, common																	
109	hay, sun-cured	1-01-645	90.0	1.26	0.15	0.41	0.31	2.03	0.24	0.32	0.063	—	—	270.0	—	—	—	31.5
110			100.0	1.40	0.17	0.45	0.35	2.26	0.27	0.35	0.070	—	—	300.0	—	—	—	35.0
	Zea mays. Corn																	
111	cobs, ground	1-02-782	90.0	0.11	—	0.06	0.04	0.79	0.42	0.42	0.117	6.6	—	208.0	5.6	—	—	0.6
112			100.0	0.12	—	0.07	0.04	0.87	0.47	0.47	0.130	7.3	—	230.0	6.2	—	—	0.7
	Ziziphus nummularia. Jujube, wild																	
113	hay, sun-cured	1-30-170	91.0	3.48	—	—	0.14	—	—	—	—	—	—	—	—	—	—	—
114			100.0	3.82	—	—	0.15	—	—	—	—	—	—	—	—	—	—	—
	PASTURE, RANGE PLANTS, AND FORAGES FED GREEN																	
	Agave spp. Agave																	
115	browse, fresh	2-00-014	12.0	0.52	—	0.04	0.01	0.09	—	—	—	—	—	—	—	—	—	—
116			100.0	4.31	—	0.29	0.11	0.74	—	—	—	—	—	—	—	—	—	—
	Agropyron desertorum. Wheatgrass, crested																	
117	fresh, early vegetative	2-05-420	28.0	0.13	—	0.08	0.10	—	—	—	—	—	—	—	—	—	—	126.0
118			100.0	0.46	—	0.28	0.34	—	—	—	—	—	—	—	—	—	—	451.0
119	fresh, post ripe	2-05-428	80.0	0.22	—	—	0.06	—	—	—	0.199	6.7	—	—	42.3	—	—	0.2
120			100.0	0.27	—	—	0.07	—	—	—	0.249	8.4	—	—	52.9	—	—	0.2
	Agropyron smithii. Wheatgrass, western																	
121	fresh, mature	2-05-407	63.0	0.31	—	0.06	0.10	0.78	—	—	0.467	8.8	—	438.0	44.2	—	23.3	39.2
122			100.0	0.50	—	0.09	0.16	1.24	—	—	0.740	14.0	—	695.0	70.0	—	37.0	62.2
	Agrostis alba. Redtop																	
123	fresh, midbloom	2-03-890	39.0	0.13	—	0.07	0.09	0.83	—	0.10	—	—	—	—	—	—	—	—
124			100.0	0.33	—	0.18	0.23	2.13	—	0.26	—	—	—	—	—	—	—	—
	Ailanthus excelsa. Ailanthus, excelsa																	
125	browse, fresh	2-30-096	26.0	0.39	—	—	0.04	—	—	—	—	—	—	—	—	—	—	—
126			100.0	1.48	—	—	0.17	—	—	—	—	—	—	—	—	—	—	—
	Ambrosia spp. Ragweed																	
127	fresh, early vegetative	2-03-851	27.0	0.45	—	—	0.04	—	—	—	—	—	—	—	—	—	—	—
128			100.0	1.68	—	—	0.16	—	—	—	—	—	—	—	—	—	—	—

TABLE 3 Composition of Goat Feeds—*Continued*

Entry Number	SCIENTIFIC NAME Feed Name	International Feed Number	Dry Matter (%)	Calcium (%)	Chlorine (%)	Magnesium (%)	Phosphorus (%)	Potassium (%)	Sodium (%)	Sulfur (%)	Cobalt (mg/kg)	Copper (mg/kg)	Iodine (mg/kg)	Iron (mg/kg)	Manganese (mg/kg)	Selenium (mg/kg)	Zinc (mg/kg)	Carotene (mg/kg)
	Andropogon barbinodis. Bluestem, cane																	
129	fresh, late vegetative	2-30-195	46.0	—	—	—	0.05	—	—	—	—	—	—	—	—	—	—	—
130			100.0	—	—	—	0.10	—	—	—	—	—	—	—	—	—	—	—
131	fresh, mature	2-30-057	89.0	—	—	—	0.03	—	—	—	—	—	—	—	—	—	—	—
132			100.0	—	—	—	0.03	—	—	—	—	—	—	—	—	—	—	—
	Andropogon scoparius. Bluestem, little																	
133	fresh, late vegetative	2-29-994	72.0	—	—	—	0.08	—	—	—	—	—	—	—	—	—	—	—
134			100.0	—	—	—	0.11	—	—	—	—	—	—	—	—	—	—	—
	Anemone heterophylla. Amemone, heterophylla																	
135	fresh, late vegetative	2-30-072	23.0	—	—	—	0.04	—	—	—	—	—	—	—	—	—	—	—
136			100.0	—	—	—	0.19	—	—	—	—	—	—	—	—	—	—	—
	Aristida wrightii. Three-awn, Wright																	
137	fresh, late vegetative	2-30-016	63.0	—	—	—	0.05	—	—	—	—	—	—	—	—	—	—	—
138			100.0	—	—	—	0.08	—	—	—	—	—	—	—	—	—	—	—
	Artemisia filifolia. Sagebrush, sand																	
139	browse, fresh, mature	2-04-135	45.0	0.22	—	—	0.05	—	—	—	—	—	—	—	—	—	—	—
140			100.0	0.48	—	—	0.12	—	—	—	—	—	—	—	—	—	—	—
	Artemisia ludoviciana albula. Sagebrush, Mexican																	
141	browse, fresh, mature	2-30-052	44.0	—	—	—	0.07	—	—	—	—	—	—	—	—	—	—	—
142			100.0	—	—	—	0.15	—	—	—	—	—	—	—	—	—	—	—
	Aster spp. Aster																	
143	fresh	2-00-443	32.0	0.47	—	0.09	0.10	0.30	—	—	—	—	—	—	13.8	—	—	—
144			100.0	1.48	—	0.29	0.31	0.94	—	—	—	—	—	—	43.0	—	—	—
	Astragalus nuttallii. Milk-vetch, Nuttall																	
145	fresh, early vegetative	2-29-814	28.0	—	—	—	0.05	—	—	—	—	—	—	—	—	—	—	—
146			100.0	—	—	—	0.18	—	—	—	—	—	—	—	—	—	—	—
	Avena sativa. Oats																	
147	fresh	2-03-292	20.0	0.05	—	0.04	0.34	0.38	—	—	—	—	—	—	—	—	—	—
148			100.0	0.27	—	0.18	1.68	1.87	—	—	—	—	—	—	—	—	—	—
	Azadirachia indica. Margosa																	
149	browse, fresh	2-30-147	21.0	0.79	—	—	0.04	—	—	—	—	—	—	—	—	—	—	—
150			100.0	3.80	—	—	0.19	—	—	—	—	—	—	—	—	—	—	—
	Bouteloua curtipendula. Grama, sideoats																	
151	fresh, late vegetative	2-29-863	62.0	—	—	—	0.07	—	—	—	—	—	—	—	—	—	—	—
152			100.0	—	—	—	0.11	—	—	—	—	—	—	—	—	—	—	—
153	aerial part without heads, fresh, mature	2-30-042	78.0	—	—	—	0.05	—	—	—	—	—	—	—	—	—	—	—
154			100.0	—	—	—	0.06	—	—	—	—	—	—	—	—	—	—	—
	Bouteloua hirsuta. Grama, hairy																	
155	fresh, late vegetative	2-30-006	55.0	—	—	—	0.05	—	—	—	—	—	—	—	—	—	—	—
156			100.0	—	—	—	0.09	—	—	—	—	—	—	—	—	—	—	—

TABLE 3 Composition of Goat Feeds—*Continued*

Entry Number	SCIENTIFIC NAME Feed Name	International Feed Number	Dry Matter (%)	Calcium (%)	Chlorine (%)	Magnesium (%)	Phosphorus (%)	Potassium (%)	Sodium (%)	Sulfur (%)	Cobalt (mg/kg)	Copper (mg/kg)	Iodine (mg/kg)	Iron (mg/kg)	Manganese (mg/kg)	Selenium (mg/kg)	Zinc (mg/kg)	Carotene (mg/kg)
	Brachiaria mutica. Paragrass																	
157	fresh	2-03-525	28.0	0.12	—	—	0.10	0.44	—	—	—	—	—	—	—	—	—	—
158			100.0	0.42	—	—	0.34	1.58	—	—	—	—	—	—	—	—	—	—
	Brassica oleracea medullosa. Kale, marrow																	
159	fresh, 71 to 84 days'	2-30-181	11.0	0.29	—	—	0.05	—	—	—	—	—	—	—	—	—	—	—
160	growth		100.0	2.59	—	—	0.43	—	—	—	—	—	—	—	—	—	—	—
161	fresh, 99 to 112 days'	2-30-182	12.0	0.37	—	—	0.05	—	—	—	—	—	—	—	—	—	—	—
162	growth		100.0	3.00	—	—	0.39	—	—	—	—	—	—	—	—	—	—	—
163	leaves, fresh	2-02-457	16.0	0.72	—	—	0.05	—	—	—	—	—	—	—	—	—	—	—
164			100.0	4.61	—	—	0.31	—	—	—	—	—	—	—	—	—	—	—
165	stems, fresh	2-08-449	14.0	0.30	—	—	0.06	—	—	—	—	—	—	—	—	—	—	—
166			100.0	2.14	—	—	0.43	—	—	—	—	—	—	—	—	—	—	—
	Bromus catharticus. Brome, rescue																	
167	fresh, late vegetative	2-08-362	33.0	—	—	—	0.06	—	—	—	—	—	—	—	—	—	—	—
168			100.0	—	—	—	0.19	—	—	—	—	—	—	—	—	—	—	—
	Buchloe dactyloides. Buffalograss																	
169	fresh, late vegetative	2-29-868	61.0	—	0.14	—	0.13	—	—	—	—	—	—	—	—	—	—	—
170			100.0	—	0.22	—	0.21	—	—	—	—	—	—	—	—	—	—	—
171	fresh, mature	2-01-008	72.0	—	—	—	—	—	—	—	—	—	—	—	—	—	—	46.4
172			100.0	—	—	—	—	—	—	—	—	—	—	—	—	—	—	64.2
	Carex planostachys. Sedge, thicket																	
173	fresh, late vegetative	2-29-865	39.0	—	—	—	0.05	—	—	—	—	—	—	—	—	—	—	—
174			100.0	—	—	—	0.13	—	—	—	—	—	—	—	—	—	—	—
	Carthamus tinctorius. Safflower																	
175	fresh	2-28-574	15.0	0.17	—	0.06	0.06	—	—	—	—	—	—	—	—	—	—	—
176			100.0	1.10	—	0.37	0.38	—	—	—	—	—	—	—	—	—	—	—
177	leaves, fresh	2-30-129	16.0	0.27	—	0.09	0.07	—	—	—	—	—	—	—	—	—	—	—
178			100.0	1.70	—	0.55	0.46	—	—	—	—	—	—	—	—	—	—	—
179	stems, fresh	2-30-130	15.0	0.12	—	—	0.04	—	—	—	—	—	—	—	—	—	—	—
180			100.0	0.80	—	—	0.28	—	—	—	—	—	—	—	—	—	—	—
	Celtis reticulata. Hackberry, netleaf																	
181	leaves, fresh	2-30-045	56.0	—	—	—	0.04	—	—	—	—	—	—	—	—	—	—	—
182			100.0	—	—	—	0.08	—	—	—	—	—	—	—	—	—	—	—
	Celtis tetendra. Hackberry, tetendra																	
183	browse, fresh, mature	2-30-164	53.0	2.10	—	—	0.07	—	—	—	—	—	—	—	—	—	—	—
184			100.0	4.00	—	—	0.13	—	—	—	—	—	—	—	—	—	—	—
	Cichorium intybus. Chicory, bluedaisy																	
185	leaves, fresh	2-19-486	23.0	0.69	—	—	0.25	—	—	—	—	—	—	—	—	—	—	—
186			100.0	2.99	—	—	1.07	—	—	—	—	—	—	—	—	—	—	—
	Commelina spp. Dayflower																	
187	fresh, late vegetative	2-30-040	17.0	—	—	—	0.02	—	—	—	—	—	—	—	—	—	—	—
188			100.0	—	—	—	0.13	—	—	—	—	—	—	—	—	—	—	—
	Croton spp. Croton																	
189	browse, fresh	2-01-694	40.0	0.85	—	0.15	0.06	0.60	—	—	—	—	—	—	—	—	—	—
190			100.0	2.12	—	0.38	0.16	1.51	—	—	—	—	—	—	—	—	—	—

TABLE 3 Composition of Goat Feeds—*Continued*

Entry Number	SCIENTIFIC NAME Feed Name	International Feed Number	Dry Matter (%)	Calcium (%)	Chlorine (%)	Magnesium (%)	Phosphorus (%)	Potassium (%)	Sodium (%)	Sulfur (%)	Cobalt (mg/kg)	Copper (mg/kg)	Iodine (mg/kg)	Iron (mg/kg)	Manganese (mg/kg)	Selenium (mg/kg)	Zinc (mg/kg)	Carotene (mg/kg)
	Cucurbita foetidissima. Buffalogourd																	
191	leaves, fresh	2-30-038	23.0	—	—	—	0.05	—	—	—	—	—	—	—	—	—	—	—
192			100.0	—	—	—	0.22	—	—	—	—	—	—	—	—	—	—	—
	Cynodon dactylon. Bermudagrass																	
193	fresh	2-00-712	34.0	0.18	—	0.06	0.07	0.57	—	—	0.025	1.9	—	—	—	—	—	104.4
194			100.0	0.53	—	0.17	0.21	1.70	—	—	0.075	5.7	—	—	—	—	—	310.3
195	fresh, late vegetative	2-10-131	38.0	—	—	—	0.08	—	—	—	—	—	—	—	—	—	—	—
196			100.0	—	—	—	0.22	—	—	—	—	—	—	—	—	—	—	—
	Dactylis glomerata. Orchardgrass																	
197	fresh, early bloom	2-03-442	25.0	0.06	—	0.08	0.10	0.84	0.01	0.06	—	8.2	—	194.0	25.8	—	—	—
198			100.0	0.25	—	0.31	0.39	3.38	0.04	0.26	—	33.1	—	785.0	104.1	—	—	—
	Dactyloctenium aegyptium. Crowfootgrass, Durban																	
199	fresh	2-21-442	17.0	0.15	—	—	0.12	—	—	—	—	—	—	—	—	—	—	—
200			100.0	0.91	—	—	0.70	—	—	—	—	—	—	—	—	—	—	—
	Dasylirion texanum. Sotol, Texas																	
201	leaves, fresh	2-04-528	57.0	0.39	—	0.06	0.05	0.43	—	—	—	—	—	—	—	—	—	—
202			100.0	0.69	—	0.11	0.08	0.76	—	—	—	—	—	—	—	—	—	—
	Digera aneensis. Digera, aneensis																	
203	fresh	2-30-149	13.0	0.37	—	—	0.07	—	—	—	—	—	—	—	—	—	—	—
204			100.0	2.91	—	—	0.57	—	—	—	—	—	—	—	—	—	—	—
	Diospyros texana. Persimmon, Texas																	
205	leaves, fresh	2-29-858	48.0	—	—	—	0.05	—	—	—	—	—	—	—	—	—	—	—
206			100.0	—	—	—	0.11	—	—	—	—	—	—	—	—	—	—	—
	Echinochloa crusgalli. Barnyardgrass																	
207	fresh	2-03-108	21.0	0.11	—	—	0.06	0.49	—	—	—	—	—	—	—	—	—	—
208			100.0	0.52	—	—	0.30	2.40	—	—	—	—	—	—	—	—	—	—
	Elymus canadensis. Wildrye, Canada																	
209	fresh, late vegetative	2-29-848	48.0	—	—	—	0.09	—	—	—	—	—	—	—	—	—	—	—
210			100.0	—	—	—	0.19	—	—	—	—	—	—	—	—	—	—	—
	Engelmannia pinnatifida. Engelmann daisy																	
211	fresh	2-29-879	30.0	—	—	—	0.05	—	—	—	—	—	—	—	—	—	—	—
212			100.0	—	—	—	0.17	—	—	—	—	—	—	—	—	—	—	—
	Eriochloa sericea. Cupgrass, Texas																	
213	fresh, late vegetative	2-29-996	40.0	—	—	—	0.06	—	—	—	—	—	—	—	—	—	—	—
214			100.0	—	—	—	0.14	—	—	—	—	—	—	—	—	—	—	—
215	fresh, mature	2-30-059	83.0	—	—	—	0.04	—	—	—	—	—	—	—	—	—	—	—
216			100.0	—	—	—	0.05	—	—	—	—	—	—	—	—	—	—	—
	Eurotia lanata. Winterfat, common																	
217	fresh, stem cured	2-26-142	80.0	1.58	—	—	0.09	—	—	—	—	—	—	—	—	—	—	14.5
218			100.0	1.98	—	—	0.12	—	—	—	—	—	—	—	—	—	—	18.1

TABLE 3 Composition of Goat Feeds—*Continued*

Entry Number	SCIENTIFIC NAME Feed Name	International Feed Number	Dry Matter (%)	Calcium (%)	Chlorine (%)	Magnesium (%)	Phosphorus (%)	Potassium (%)	Sodium (%)	Sulfur (%)	Cobalt (mg/kg)	Copper (mg/kg)	Iodine (mg/kg)	Iron (mg/kg)	Manganese (mg/kg)	Selenium (mg/kg)	Zinc (mg/kg)	Carotene (mg/kg)
	Evax prolifera. Evax, prolifera																	
219	fresh, early vegetative	2-30-068	38.0	—	—	—	0.08	—	—	—	—	—	—	—	—	—	—	—
220			100.0	—	—	—	0.20	—	—	—	—	—	—	—	—	—	—	—
	Ficus benghalensis. Banyan-tree																	
221	leaves, fresh	2-27-208	32.0	0.81	—	—	0.13	—	—	—	—	—	—	—	—	—	—	—
222			100.0	2.53	—	—	0.40	—	—	—	—	—	—	—	—	—	—	—
	Ficus glomerata. Fig, cluster																	
223	leaves, fresh	2-27-209	30.0	1.13	—	—	0.21	—	—	—	—	—	—	—	—	—	—	—
224			100.0	3.75	—	—	0.71	—	—	—	—	—	—	—	—	—	—	—
	Ficus roxburghii. Fig, Roxburgh																	
225	fresh	2-30-177	33.0	0.44	—	—	0.06	—	—	—	—	—	—	—	—	—	—	—
226			100.0	1.31	—	—	0.17	—	—	—	—	—	—	—	—	—	—	—
	Forestiera pubescens. Forestiera, downy																	
227	browse, fresh, late vegetative	2-29-837	38.0	—	—	—	0.06	—	—	—	—	—	—	—	—	—	—	—
228			100.0	—	—	—	0.16	—	—	—	—	—	—	—	—	—	—	—
	Grewia oppositifolia roxburghii. Grewia, oppositifolia, roxburghii																	
229	browse, fresh	2-30-097	33.0	0.90	—	0.18	0.08	—	—	—	—	—	—	—	—	—	—	—
230			100.0	2.74	—	0.54	0.24	—	—	—	—	—	—	—	—	—	—	—
	Gutierrezia texana. Snakeweed, Texas																	
231	fresh, late vegetative	2-04-245	30.0	0.20	—	0.06	0.05	0.89	—	—	—	—	—	—	—	—	—	—
232			100.0	0.65	—	0.21	0.17	2.98	—	—	—	—	—	—	—	—	—	—
	Hilaria belangeri. Curly mesquite																	
233	browse, fresh, early vegetative	2-01-723	23.0	0.24	—	0.07	0.06	0.18	—	—	—	—	—	—	—	—	—	—
234			100.0	1.04	—	0.31	0.26	0.79	—	—	—	—	—	—	—	—	—	—
	Hordeum vulgare. Barley																	
235	fresh	2-00-511	21.0	0.13	—	—	0.09	—	—	—	—	—	—	—	—	—	—	49.1
236			100.0	0.60	—	—	0.40	—	—	—	—	—	—	—	—	—	—	230.4
	Juniperus ashei. Juniper, ashe																	
237	leaves, fresh	2-29-850	52.0	—	—	—	0.04	—	—	—	—	—	—	—	—	—	—	—
238			100.0	—	—	—	0.08	—	—	—	—	—	—	—	—	—	—	—
	Juniperus pinchotii. Juniper, redberry																	
239	leaves, fresh	2-29-851	53.0	—	—	—	0.06	—	—	—	—	—	—	—	—	—	—	—
240			100.0	—	—	—	0.11	—	—	—	—	—	—	—	—	—	—	—
	Lactuca sativa. Lettuce																	
241	fresh	2-02-624	5.0	0.05	—	—	0.03	0.24	0.01	—	—	—	—	13.0	—	—	—	—
242			100.0	0.86	—	—	0.46	4.52	0.17	—	—	—	—	246.0	—	—	—	—
	Lepidium spp. Pepperweed																	
243	fresh, late vegetative	2-03-686	42.0	0.83	—	0.21	0.17	1.35	—	—	—	—	—	—	—	—	—	—
244			100.0	1.95	—	0.49	0.41	3.18	—	—	—	—	—	—	—	—	—	—

TABLE 3 Composition of Goat Feeds—*Continued*

Entry Number	SCIENTIFIC NAME Feed Name	International Feed Number	Dry Matter (%)	Calcium (%)	Chlorine (%)	Magnesium (%)	Phosphorus (%)	Potassium (%)	Sodium (%)	Sulfur (%)	Cobalt (mg/kg)	Copper (mg/kg)	Iodine (mg/kg)	Iron (mg/kg)	Manganese (mg/kg)	Selenium (mg/kg)	Zinc (mg/kg)	Carotene (mg/kg)
	Leptochloa dubia. Sprangletop, green																	
245	fresh, late vegetative	2-30-031	52.0	—	—	—	0.05	—	—	—	—	—	—	—	—	—	—	—
246			100.0	—	—	—	0.09	—	—	—	—	—	—	—	—	—	—	—
	Leptoloma cognatum. Witchgrass, fall																	
247	fresh, late vegetative	2-30-039	50.0	—	—	—	0.04	—	—	—	—	—	—	—	—	—	—	—
248			100.0	—	—	—	0.07	—	—	—	—	—	—	—	—	—	—	—
	Lespedeza striata. Lespedeza, common																	
249	fresh, full bloom	2-02-541	35.0	0.42	—	0.09	0.09	0.35	—	—	—	—	—	94.0	61.4	—	—	—
250			100.0	1.21	—	0.25	0.27	1.01	—	—	—	—	—	270.0	176.4	—	—	—
	Lesquerella gordonii. Bladderpod, Gordon																	
251	fresh, late vegetative	2-29-820	35.0	—	—	—	0.06	—	—	—	—	—	—	—	—	—	—	—
252			100.0	—	—	—	0.17	—	—	—	—	—	—	—	—	—	—	—
	Lonicera albiflora. Honeysuckle, bush																	
253	leaves, fresh	2-29-875	33.0	—	—	—	0.04	—	—	—	—	—	—	—	—	—	—	—
254			100.0	—	—	—	0.11	—	—	—	—	—	—	—	—	—	—	—
	Lotus spp. Trefoil																	
255	fresh	2-01-757	15.0	0.32	—	—	0.04	0.32	—	—	—	—	—	—	—	—	—	—
256			100.0	2.07	—	—	0.28	2.09	—	—	—	—	—	—	—	—	—	—
	Lupinus texensis. Bluebonnet, Texas																	
257	fresh, early vegetative	2-29-821	18.0	—	—	—	0.02	—	—	—	—	—	—	—	—	—	—	—
258			100.0	—	—	—	0.14	—	—	—	—	—	—	—	—	—	—	—
259	fresh, late vegetative	2-29-831	16.0	—	—	—	0.03	—	—	—	—	—	—	—	—	—	—	—
260			100.0	—	—	—	0.17	—	—	—	—	—	—	—	—	—	—	—
	Medicago sativa. Alfalfa																	
261	fresh	2-00-196	24.0	0.48	0.11	0.07	0.07	0.51	0.05	0.09	0.032	2.4	—	70.0	10.4	—	4.3	45.0
262			100.0	1.96	0.47	0.27	0.30	2.09	0.19	0.37	0.133	9.9	—	286.0	42.6	—	17.6	185.0
263	fresh, late vegetative	2-00-181	21.0	0.47	0.09	0.06	0.07	0.46	0.05	0.10	0.036	2.3	—	24.0	8.7	—	—	—
264			100.0	2.19	0.44	0.27	0.33	2.14	0.21	0.48	0.167	10.8	—	111.0	40.8	—	—	—
265	fresh, full bloom	2-00-188	25.0	0.38	0.11	0.07	0.07	0.53	0.04	0.08	—	—	—	107.0	38.7	—	—	—
266			100.0	1.53	0.43	0.27	0.27	2.13	0.15	0.31	—	—	—	430.0	155.2	—	—	—
	Mimosa spp. Mimosa																	
267	browse, fresh, early	2-03-122	26.0	0.62	—	0.08	0.04	0.30	—	—	—	—	—	—	—	—	—	—
268	vegetative		100.0	2.38	—	0.32	0.17	1.14	—	—	—	—	—	—	—	—	—	—
	Morus australis. Mulberry, Japanese																	
269	leaves, fresh	2-30-179	15.0	0.36	—	—	0.04	—	—	—	—	—	—	—	—	—	—	—
270			100.0	2.42	—	—	0.24	—	—	—	—	—	—	—	—	—	—	—
	Musa spp. Banana																	
271	leaves, fresh	2-09-902	19.0	0.23	—	0.06	0.13	—	—	—	—	—	—	—	—	—	—	—
272			100.0	1.20	—	0.32	0.70	—	—	—	—	—	—	—	—	—	—	—
	Nolina texana. Nolina, Texas																	
273	buds, fresh	2-29-855	32.0	—	—	—	0.12	—	—	—	—	—	—	—	—	—	—	—
274			100.0	—	—	—	0.38	—	—	—	—	—	—	—	—	—	—	—
275	leaves, fresh	2-29-852	57.0	—	—	—	0.04	—	—	—	—	—	—	—	—	—	—	—
276			100.0	—	—	—	0.06	—	—	—	—	—	—	—	—	—	—	—

TABLE 3 Composition of Goat Feeds—*Continued*

Entry Number	SCIENTIFIC NAME Feed Name	International Feed Number	Dry Matter (%)	Calcium (%)	Chlorine (%)	Magnesium (%)	Phosphorus (%)	Potassium (%)	Sodium (%)	Sulfur (%)	Cobalt (mg/kg)	Copper (mg/kg)	Iodine (mg/kg)	Iron (mg/kg)	Manganese (mg/kg)	Selenium (mg/kg)	Zinc (mg/kg)	Carotene (mg/kg)
	Opuntia spp. Pricklypear																	
277	fresh, mature	2-01-059	21.0	—	0.05	—	0.01	—	0.06	0.05	—	—	—	—	—	—	—	1.3
278			100.0	—	0.21	—	0.03	—	0.30	0.23	—	—	—	—	—	—	—	6.0
279	fruit, fresh, immature	4-30-020	26.0	—	—	—	0.03	—	—	—	—	—	—	—	—	—	—	—
280			100.0	—	—	—	0.13	—	—	—	—	—	—	—	—	—	—	—
	Panicum hallii. Panicum, Halls																	
281	fresh, late vegetative	2-29-998	49.0	—	—	—	0.05	—	—	—	—	—	—	—	—	—	—	—
282			100.0	—	—	—	0.10	—	—	—	—	—	—	—	—	—	—	—
	Panicum maximum. Guineagrass																	
283	fresh, 15 to 28 days'	2-09-995	16.0	0.09	—	0.05	0.06	—	—	—	—	—	—	—	—	—	—	—
284	growth		100.0	0.54	—	0.30	0.36	—	—	—	—	—	—	—	—	—	—	—
285	fresh, 29 to 42 days'	2-09-714	18.0	0.10	—	0.05	0.06	—	—	—	—	—	—	—	—	—	—	—
286	growth		100.0	0.53	—	0.30	0.32	—	—	—	—	—	—	—	—	—	—	—
287	fresh, 43 to 56 days'	2-09-910	21.0	0.10	—	0.04	0.07	—	—	—	—	—	—	—	—	—	—	—
288	growth		100.0	0.51	—	0.19	0.33	—	—	—	—	—	—	—	—	—	—	—
	Panicum obtusum. Vine-mesquite																	
289	fresh, late vegetative	2-30-009	47.0	—	—	—	0.07	—	—	—	—	—	—	—	—	—	—	—
290			100.0	—	—	—	0.14	—	—	—	—	—	—	—	—	—	—	—
	Panicum virgatum. Switch-grass																	
291	fresh, early vegetative	2-04-795	35.0	0.16	—	—	0.07	—	—	—	—	—	—	—	—	—	—	—
292			100.0	0.46	—	—	0.20	—	—	—	—	—	—	—	—	—	—	—
	Paspalum dilatatum. Dallisgrass																	
293	fresh, early vegetative	2-01-738	26.0	0.17	—	—	0.11	—	—	—	—	—	—	—	—	—	—	108.8
294			100.0	0.65	—	—	0.42	—	—	—	—	—	—	—	—	—	—	426.6
	Pennisetum incomptum. Pennisetum, incomptum																	
295	fresh	2-30-104	30.0	0.19	—	0.16	0.14	—	—	—	—	—	—	—	—	—	—	—
296			100.0	0.63	—	0.53	0.45	—	—	—	—	—	—	—	—	—	—	—
	Pennisetum purpureum. Napiergrass																	
297	fresh	2-03-166	21.0	0.09	—	0.06	0.07	0.28	0.00	0.02	—	—	—	—	—	—	—	—
298			100.0	0.44	—	0.26	0.35	1.31	0.01	0.10	—	—	—	—	—	—	—	—
299	fresh, 15 to 28 days'	2-09-412	15.0	0.05	—	0.07	0.03	—	—	—	—	—	—	—	—	—	—	—
300	growth		100.0	0.35	—	0.48	0.23	—	—	—	—	—	—	—	—	—	—	—
301	fresh, 29 to 42 days'	2-10-095	21.0	0.08	—	0.08	0.06	—	—	—	—	—	—	—	—	—	—	—
302	growth		100.0	0.39	—	0.37	0.31	—	—	—	—	—	—	—	—	—	—	—
	Phalaris arundinacea. Canarygrass, reed																	
303	fresh	2-01-113	27.0	0.11	—	—	0.09	0.97	—	—	—	—	—	—	—	—	—	—
304			100.0	0.41	—	—	0.35	3.64	—	—	—	—	—	—	—	—	—	—
	Phalaris tuberosa stenoptera. Hardinggrass																	
305	fresh	2-02-354	27.0	0.08	—	—	0.03	—	—	—	—	—	—	—	—	—	—	—
306			100.0	0.29	—	—	0.12	—	—	—	—	—	—	—	—	—	—	—
	Plantago rhodosperma. Plantain, redseed																	
307	leaves, fresh	2-29-870	28.0	—	—	—	0.03	—	—	—	—	—	—	—	—	—	—	—
308			100.0	—	—	—	0.10	—	—	—	—	—	—	—	—	—	—	—

TABLE 3 Composition of Goat Feeds—*Continued*

Entry Number	SCIENTIFIC NAME Feed Name	International Feed Number	Dry Matter (%)	Calcium (%)	Chlorine (%)	Magnesium (%)	Phosphorus (%)	Potassium (%)	Sodium (%)	Sulfur (%)	Cobalt (mg/kg)	Copper (mg/kg)	Iodine (mg/kg)	Iron (mg/kg)	Manganese (mg/kg)	Selenium (mg/kg)	Zinc (mg/kg)	Carotene (mg/kg)
	Prosopis glandulosa. Mesquite, honey																	
309	browse, fresh, late	2-29-999	48.0	—	—	—	0.04	—	—	—	—	—	—	—	—	—	—	—
310	vegetative		100.0	—	—	—	0.08	—	—	—	—	—	—	—	—	—	—	—
	Quercus sinuata brevi-loba. Oak, shin																	
311	browse, fresh, early	2-29-826	32.0	—	—	—	0.10	—	—	—	—	—	—	—	—	—	—	—
312	vegetative		100.0	—	—	—	0.31	—	—	—	—	—	—	—	—	—	—	—
	Quercus virginiana. Oak, live																	
313	browse, fresh, early	2-29-815	36.0	—	—	—	0.14	—	—	—	—	—	—	—	—	—	—	—
314	vegetative		100.0	—	—	—	0.38	—	—	—	—	—	—	—	—	—	—	—
315	leaves, fresh	2-29-859	50.0	—	—	—	0.05	—	—	—	—	—	—	—	—	—	—	—
316			100.0	—	—	—	0.11	—	—	—	—	—	—	—	—	—	—	—
	Ratibida columnifera. Prairieconeflower, upright																	
317	fresh, early vegetative	2-29-846	18.0	—	—	—	0.06	—	—	—	—	—	—	—	—	—	—	—
318			100.0	—	—	—	0.33	—	—	—	—	—	—	—	—	—	—	—
	Rhus aromatica. Sumac, fragrant																	
319	browse, fresh, early	2-29-827	40.0	—	—	—	0.08	—	—	—	—	—	—	—	—	—	—	—
320	vegetative		100.0	—	—	—	0.20	—	—	—	—	—	—	—	—	—	—	—
	Sarcobatus vermiculatus. Greasewood, black																	
321	browse, fresh	2-20-083	35.0	0.39	—	0.17	0.07	0.73	—	0.18	0.098	5.9	—	—	32.7	—	11.6	—
322			100.0	1.10	—	0.48	0.21	2.09	—	0.50	0.280	16.9	—	—	93.4	—	33.1	—
	Schrankia roemeriana. Sensitivebrier, Roemer																	
323	browse, fresh, late	2-29-860	32.0	—	—	—	0.07	—	—	—	—	—	—	—	—	—	—	—
324	vegetative		100.0	—	—	—	0.22	—	—	—	—	—	—	—	—	—	—	—
	Secale cereale. Rye																	
325	fresh	2-04-018	24.0	0.09	—	0.08	0.08	0.82	0.02	—	—	—	—	—	—	—	—	82.3
326			100.0	0.39	—	0.31	0.33	3.40	0.07	—	—	—	—	—	—	—	—	342.6
	Sedum nuttallianum. Stonecrop, Nuttall																	
327	fresh, early vegetative	2-29-836	10.0	—	—	—	0.02	—	—	—	—	—	—	—	—	—	—	—
328			100.0	—	—	—	0.20	—	—	—	—	—	—	—	—	—	—	—
	Setaria macrostachya. Bristlegrass, plains																	
329	fresh, late vegetative	2-29-882	43.0	—	—	—	0.09	—	—	—	—	—	—	—	—	—	—	—
330			100.0	—	—	—	0.20	—	—	—	—	—	—	—	—	—	—	—
	Sorghum bicolor sudanense. Sorghum, Sudangrass																	
331	fresh, early vegetative	2-04-484	18.0	0.08	—	0.06	0.07	0.38	0.00	0.02	—	—	—	36.0	—	—	—	35.2
332			100.0	0.43	—	0.35	0.41	2.14	0.01	0.11	—	—	—	200.0	—	—	—	197.5
333	fresh, midbloom	2-04-485	23.0	0.10	—	0.08	0.08	0.49	0.00	0.03	—	—	—	46.0	—	—	—	41.6
334			100.0	0.43	—	0.35	0.36	2.14	0.01	0.11	—	—	—	200.0	—	—	—	182.5
	Sorghum halepense. Sorghum, Johnsongrass																	
335	fresh	2-04-412	24.0	0.18	—	0.06	0.05	0.74	—	—	—	—	—	—	—	—	—	47.0
336			100.0	0.77	—	0.25	0.23	3.12	—	—	—	—	—	—	—	—	—	198.4

TABLE 3 Composition of Goat Feeds—*Continued*

Entry Number	SCIENTIFIC NAME Feed Name	International Feed Number	Dry Matter (%)	Calcium (%)	Chlorine (%)	Magnesium (%)	Phosphorus (%)	Potassium (%)	Sodium (%)	Sulfur (%)	Cobalt (mg/kg)	Copper (mg/kg)	Iodine (mg/kg)	Iron (mg/kg)	Manganese (mg/kg)	Selenium (mg/kg)	Zinc (mg/kg)	Carotene (mg/kg)
	Stipa leucotricha. Needlegrass, Texas																	
337	fresh, late vegetative	2-29-822	55.0	—	—	—	0.06	—	—	—	—	—	—	—	—	—	—	—
338			100.0	—	—	—	0.10	—	—	—	—	—	—	—	—	—	—	—
	Tillandsia usneoides. Spanishmoss																	
339	whole, fresh	2-04-648	41.0	—	—	—	—	—	—	—	—	—	—	—	—	—	—	6.2
340			100.0	—	—	—	—	—	—	—	—	—	—	—	—	—	—	15.0
	Tribulus terrestris. Puncturevine																	
341	fresh	2-27-107	22.0	0.77	—	—	0.06	—	—	—	—	—	—	—	—	—	—	—
342			100.0	3.50	—	—	0.26	—	—	—	—	—	—	—	—	—	—	—
	Trifolium alexandrinum. Clover, Egyptian																	
343	fresh	2-01-349	18.0	0.63	—	—	0.06	0.41	—	—	—	—	—	—	—	—	—	—
344			100.0	3.56	—	—	0.32	2.31	—	—	—	—	—	—	—	—	—	—
	Triodia albescens. Triodia, white																	
345	fresh, late vegetative	2-30-011	43.0	—	—	—	0.06	—	—	—	—	—	—	—	—	—	—	—
346			100.0	—	—	—	0.15	—	—	—	—	—	—	—	—	—	—	—
	Vicia sativa. Vetch, common																	
347	fresh	2-05-123	20.0	0.27	—	0.04	0.07	0.51	—	0.02	—	2.0	—	80.0	24.5	—	—	—
348			100.0	1.32	—	0.20	0.34	2.50	—	0.10	—	9.7	—	392.0	120.0	—	—	—
	Vigna sinensis. Cowpea, common																	
349	fresh	2-01-655	15.0	0.29	0.03	0.07	0.04	0.25	0.04	0.05	—	—	—	120.0	—	—	—	—
350			100.0	1.91	0.18	0.43	0.28	1.66	0.25	0.31	—	—	—	798.0	—	—	—	—
	Yucca spp. Yucca																	
351	flowers, fresh	2-05-536	15.0	—	—	—	0.07	—	—	—	—	—	—	—	—	—	—	—
352			100.0	—	—	—	0.48	—	—	—	—	—	—	—	—	—	—	—
353	leaves, fresh, immature	2-30-050	41.0	—	—	—	0.04	—	—	—	—	—	—	—	—	—	—	—
354			100.0	—	—	—	0.10	—	—	—	—	—	—	—	—	—	—	—
	Zexmenia hispida. Zexmenia, orange																	
355	leaves, fresh	2-30-051	39.0	—	—	—	0.04	—	—	—	—	—	—	—	—	—	—	—
356			100.0	—	—	—	0.10	—	—	—	—	—	—	—	—	—	—	—
	SILAGES																	
	Avena sativa. Oats																	
357	silage	3-03-298	31.0	0.10	—	0.09	0.07	0.84	0.07	0.09	0.017	1.7	—	65.0	13.3	0.0	10.7	13.7
358			100.0	0.34	—	0.30	0.24	2.74	0.23	0.29	0.057	5.5	—	211.0	43.3	0.1	35.0	44.6
	Beta vulgaris altissima. Beet, sugar																	
359	aerial part with crowns,	3-00-660	22.0	0.35	—	0.24	0.06	1.28	0.12	0.13	—	—	—	45.0	—	—	—	—
360	silage		100.0	1.56	—	1.07	0.29	5.74	0.54	0.57	—	—	—	200.0	—	—	—	—
	Digitaria decumbens. Pangolagrass																	
361	silage, mature	3-03-494	32.0	0.13	—	0.01	0.03	—	—	—	—	—	—	—	—	—	—	—
362			100.0	0.39	—	0.03	0.08	—	—	—	—	—	—	—	—	—	—	—
	Glycine max. Soybean																	
363	silage wilted	3-04-584	35.0	0.48	—	0.16	0.12	0.32	0.03	0.10	—	3.2	—	62.0	45.4	—	—	—
364			100.0	1.38	—	0.45	0.36	0.93	0.09	0.30	—	9.2	—	179.0	131.1	—	—	—

TABLE 3 Composition of Goat Feeds—*Continued*

Entry Number	SCIENTIFIC NAME Feed Name	International Feed Number	Dry Matter (%)	Calcium (%)	Chlorine (%)	Magnesium (%)	Phosphorus (%)	Potassium (%)	Sodium (%)	Sulfur (%)	Cobalt (mg/kg)	Copper (mg/kg)	Iodine (mg/kg)	Iron (mg/kg)	Manganese (mg/kg)	Selenium (mg/kg)	Zinc (mg/kg)	Carotene (mg/kg)
	Medicago sativa. Alfalfa																	
365	silage, full bloom	3-00-207	26.0	—	—	—	—	—	—	—	—	—	—	—	—	—	—	10.8
366			100.0	—	—	—	—	—	—	—	—	—	—	—	—	—	—	41.0
	Pennisetum purpureum. Napiergrass																	
367	silage	3-03-170	26.0	0.08	—	—	0.08	—	—	—	—	—	—	—	—	—	—	24.7
368			100.0	0.31	—	—	0.31	—	—	—	—	—	—	—	—	—	—	96.6
	Pisum spp. Pea																	
369	silage	3-03-590	24.0	0.62	—	—	0.12	—	—	—	—	—	—	—	—	—	—	31.0
370			100.0	2.56	—	—	0.48	—	—	—	—	—	—	—	—	—	—	128.7
	Sorghum bicolor saccharatum. Sorghum, sorgo																	
371	silage, dough stage	3-04-466	26.0	0.07	—	—	0.04	—	—	—	—	—	—	—	—	—	—	—
372			100.0	0.27	—	—	0.15	—	—	—	—	—	—	—	—	—	—	—
373	silage, mature	3-04-467	27.0	0.07	—	—	0.04	—	—	—	—	—	—	—	—	—	—	0.8
374			100.0	26.0	—	—	0.14	—	—	—	—	—	—	—	—	—	—	2.9
	Zea mays. Corn																	
375	silage, milk stage	3-02-818	22.0	0.09	—	0.09	0.07	0.35	0.00	—	—	—	—	—	—	—	—	—
376			100.0	0.41	—	0.41	0.29	1.57	0.01	—	—	—	—	—	—	—	—	—
377	silage, dough stage	3-02-819	26.0	0.07	—	0.05	0.05	0.25	0.00	0.04	—	—	—	48.0	—	—	—	17.2
378			100.0	0.27	—	0.18	0.19	0.95	0.01	0.14	—	—	—	180.0	—	—	—	65.1
	ENERGY FEEDS																	
	Avena sativa. Oats																	
379	grain	4-03-309	89.0	0.07	0.09	0.13	0.33	0.39	0.07	0.21	0.055	6.2	0.10	76.0	36.9	0.2	36.9	0.1
380			100.0	0.07	0.11	0.14	0.38	0.44	0.08	0.23	0.062	7.0	0.11	85.0	41.5	0.3	41.5	0.1
	Beta vulgaris altissima. Beet, sugar																	
381	molasses, more than 48%	4-00-668	78.0	0.13	1.28	0.23	0.03	4.72	1.15	0.46	0.361	16.8	—	68.0	4.5	—	14.0	—
382	invert sugar, more than 79.5 degrees brix		100.0	0.17	1.64	0.29	0.03	6.07	1.48	0.60	0.464	21.6	—	87.0	5.8	—	18.0	—
383	pulp, dehydrated	4-00-669	91.0	0.63	0.04	0.24	0.09	0.18	0.19	0.20	0.073	12.5	—	299.0	34.6	—	0.7	0.2
384			100.0	0.69	0.04	0.27	0.10	0.20	0.21	0.22	0.081	13.8	—	329.0	38.2	—	0.8	0.2
385	roots, fresh	4-00-677	20.0	0.05	0.10	0.04	0.05	0.30	0.10	0.01	—	1.6	—	12.0	40.3	—	—	—
386			100.0	0.24	0.49	0.18	0.24	1.52	0.49	0.06	—	8.1	—	61.0	205.7	—	—	—
	Bos taurus. Cattle																	
387	whey, dehydrated	4-01-182	93.0	0.86	0.07	0.13	0.76	1.15	0.65	1.04	0.110	46.6	—	169.0	5.9	—	3.2	—
388			100.0	0.92	0.08	0.14	0.82	1.23	0.70	1.12	0.118	49.9	—	181.0	6.3	—	3.4	—
	Brassica oleracea capitata. Cabbage																	
389	fresh	4-01-046	9.0	0.06	0.05	0.02	0.03	0.26	0.02	0.11	—	1.3	—	7.0	2.8	—	—	0.4
390			100.0	0.64	0.53	0.21	0.35	2.81	0.19	1.17	—	14.1	—	80.0	30.5	—	—	4.7
	Brassica rapa rapa. Turnip																	
391	roots, fresh	4-05-067	9.0	0.05	0.06	0.02	0.02	0.28	0.10	0.04	—	2.0	—	10.0	4.0	—	—	—
392			100.0	0.59	0.65	0.22	0.26	2.99	1.05	0.43	—	21.3	—	112.0	42.7	—	—	—
	Ceratonia siliqua. Carob bean																	
393	pods with seeds	4-08-370	88.0	0.20	—	—	0.30	—	—	—	—	—	—	—	—	—	—	—
394			100.0	0.23	—	—	0.34	—	—	—	—	—	—	—	—	—	—	—
	Citrus spp. Citrus																	
395	pomace, dehydrated or	4-08-652	90.0	1.79	—	—	0.12	0.62	0.05	—	0.220	5.7	—	163.0	6.8	—	14.5	—
396	sun-cured (pulp)		100.0	1.98	—	—	0.13	0.69	0.06	—	0.244	6.4	—	180.0	7.6	—	16.1	—

TABLE 3 Composition of Goat Feeds—*Continued*

Entry Number	SCIENTIFIC NAME Feed Name	International Feed Number	Dry Matter (%)	Calcium (%)	Chlorine (%)	Magnesium (%)	Phosphorus (%)	Potassium (%)	Sodium (%)	Sulfur (%)	Cobalt (mg/kg)	Copper (mg/kg)	Iodine (mg/kg)	Iron (mg/kg)	Manganese (mg/kg)	Selenium (mg/kg)	Zinc (mg/kg)	Carotene (mg/kg)
	Daucus spp. Carrot																	
397	roots, fresh	4-01-145	12.0	0.05	0.06	0.02	0.04	0.33	0.12	0.02	—	1.2	—	14.0	3.7	—	—	80.1
398			100.0	0.40	0.50	0.20	0.35	2.80	1.04	0.17	—	10.4	—	120.0	31.5	—	—	677.5
	Hordeum vulgare. Barley																	
399	grain	4-00-549	88.0	0.04	0.16	0.14	0.34	0.41	0.03	0.15	0.087	7.9	0.04	75.0	16.1	0.2	16.9	2.0
400			100.0	0.05	0.18	0.15	0.38	0.47	0.03	0.17	0.099	9.0	0.05	85.0	18.3	0.2	19.1	2.3
	Ipomoea batatas. Sweet-potato																	
401	tubers, fresh	4-04-788	33.0	0.03	0.02	0.05	0.05	0.34	0.02	0.04	—	1.4	—	16.0	3.7	—	—	143.8
402			100.0	0.10	0.06	0.16	0.15	1.01	0.05	0.13	—	4.2	—	48.0	11.1	—	—	431.2
	Laminariales (order)-fucales (order). Seaweed, kelp																	
403	whole, sun-cured	4-04-190	89.0	1.32	—	5.68	0.16	—	—	—	—	—	—	250.0	—	—	—	—
404			100.0	1.48	—	6.37	0.18	—	—	—	—	—	—	281.0	—	—	—	—
	Malus spp. Apple																	
405	pomace, wet	4-00-424	20.0	0.02	—	—	0.02	0.10	—	—	—	—	—	—	—	—	—	—
406			100.0	0.10	—	—	0.10	0.47	—	—	—	—	—	—	—	—	—	—
	Oryza sativa. Rice																	
407	bran with germs	4-03-928	91.0	0.07	0.07	0.94	1.54	1.74	0.03	0.18	—	13.2	—	190.0	376.0	0.4	29.4	—
408			100.0	0.08	0.08	1.04	1.70	1.92	0.04	0.20	—	14.5	—	210.0	414.8	0.4	32.4	—
409	grain	4-03-939	89.0	0.06	0.08	0.13	0.27	0.32	0.04	0.04	0.044	3.1	0.04	51.0	17.6	—	14.7	—
410			100.0	0.07	0.09	0.15	0.30	0.36	0.04	0.05	0.050	3.5	0.05	57.0	19.8	—	16.6	—
	Pennisetum glaucum. Pearl-millet																	
411	grain	4-03-118	90.0	0.05	0.14	0.16	0.31	0.43	0.04	0.13	0.044	21.9	—	24.0	30.7	—	13.2	—
412			100.0	0.06	0.16	0.18	0.35	0.48	0.04	0.14	0.049	24.5	—	27.0	34.3	—	14.7	—
	Saccharum officinarum. Sugarcane																	
413	molasses, more than 46%	4-04-696	75.0	0.75	2.31	0.32	0.08	2.86	0.16	0.35	0.902	59.2	1.57	186.0	42.0	—	22.4	—
414	invert sugars, more than 79.5 degrees brix		100.0	1.00	3.10	0.43	0.11	3.84	0.22	0.47	1.210	79.4	2.10	250.0	56.3	—	30.0	—
	Secale cereale. Rye																	
415	grain	4-04-047	88.0	0.06	0.03	0.12	0.32	0.46	0.02	0.15	—	6.8	—	60.0	57.6	0.4	31.3	0.1
416			100.0	0.07	0.03	0.14	0.37	0.52	0.03	0.17	—	7.8	—	69.0	65.8	0.4	35.7	0.1
	Solanum tuberosum. Potato																	
417	tubers, fresh	4-03-787	23.0	0.01	0.07	0.03	0.06	0.51	0.02	0.02	—	6.6	—	18.0	9.8	—	—	—
418			100.0	0.04	0.28	0.14	0.24	2.17	0.09	0.09	—	28.4	—	78.0	41.7	—	—	—
419	tubers, silage	4-03-768	25.0	0.01	—	0.04	0.06	0.53	0.02	0.06	—	—	—	22.0	—	—	—	—
420			100.0	0.04	—	0.14	0.23	2.13	0.09	0.23	—	—	—	90.0	—	—	—	—
	Sorghum bicolor subglabrescens. Sorghum, milo																	
421	grain	4-04-444	89.0	0.03	0.08	0.16	0.28	0.34	0.02	0.11	0.474	11.0	0.06	50.0	17.3	0.1	21.9	0.2
422			100.0	0.04	0.09	0.18	0.32	0.38	0.02	0.13	0.531	12.4	0.07	56.0	19.4	0.1	24.5	0.3
	Triticum aestivum. Wheat																	
423	bran	4-05-190	89.0	0.11	0.05	0.53	1.22	1.38	0.04	0.22	0.100	12.7	0.07	114.0	110.6	0.4	113.7	2.6
424			100.0	0.13	0.05	0.60	1.38	1.56	0.04	0.25	0.113	14.3	0.07	128.0	124.5	0.4	128.0	2.9
425	grain	4-05-211	89.0	0.04	0.07	0.15	0.37	0.38	0.04	0.16	0.120	6.3	0.09	54.0	37.0	0.3	44.2	0.1
426			100.0	0.04	0.08	0.16	0.42	0.42	0.05	0.18	0.135	7.1	0.10	61.0	41.6	0.3	49.7	0.1
	Zea mays. Corn																	
427	ears, ground	4-02-849	87.0	0.06	0.04	0.12	0.24	0.46	0.02	0.14	0.272	6.8	0.02	79.0	12.3	0.1	12.1	3.2
428			100.0	0.07	0.05	0.14	0.27	0.53	0.02	0.16	0.314	7.9	0.03	91.0	14.2	0.1	14.0	3.7

TABLE 3 Composition of Goat Feeds—*Continued*

Entry Number	SCIENTIFIC NAME Feed Name	International Feed Number	Dry Matter (%)	Calcium (%)	Chlorine (%)	Magnesium (%)	Phosphorus (%)	Potassium (%)	Sodium (%)	Sulfur (%)	Cobalt (mg/kg)	Copper (mg/kg)	Iodine (mg/kg)	Iron (mg/kg)	Manganese (mg/kg)	Selenium (mg/kg)	Zinc (mg/kg)	Carotene (mg/kg)
429	grain	4-02-879	87.0	0.04	—	0.10	0.24	0.96	0.01	0.12	0.019	2.1	—	43.0	4.9	0.3	19.8	—
430			100.0	0.05	—	0.11	0.28	1.10	0.02	0.14	0.022	2.4	—	50.0	5.6	0.3	22.7	—
	Zea mays indentata. Corn, dent yellow																	
431	grain	4-02-935	89.0	0.03	0.04	0.12	0.26	0.33	0.03	0.11	0.046	3.5	—	27.0	4.8	0.1	12.8	2.2
432			100.0	0.03	0.05	0.14	0.29	0.37	0.03	0.12	0.052	4.0	—	30.0	5.4	0.1	14.4	2.5
	PROTEIN SUPPLEMENTS																	
	Arachis hypogaea. Peanut																	
433	kernels	5-03-657	95.0	0.06	0.02	0.18	0.43	0.61	0.29	0.25	—	—	—	20.0	—	—	—	0.2
434			100.0	0.06	0.02	0.19	0.45	0.64	0.30	0.26	—	—	—	21.0	—	—	—	0.2
435	kernels, meal solvent extracted	5-03-650	92.0	0.26	0.03	0.15	0.62	1.13	0.07	0.30	0.109	15.3	0.07	142.0	26.8	—	19.9	—
436			100.0	0.29	0.03	0.17	0.68	1.23	0.08	0.33	0.119	16.6	0.07	154.0	29.2	—	21.7	—
	Blood, meal spray dehy																	
437	(blood flour)	5-00-381	93.0	0.48	0.25	0.22	0.24	0.09	0.39	0.34	—	8.2	—	2784.0	6.4	—	—	—
438			100.0	0.52	0.27	0.24	0.26	0.10	0.42	0.37	—	8.8	—	2993.0	6.9	—	—	—
	Bos taurus. Cattle																	
439	milk, fresh	5-01-168	12.0	0.12	0.11	0.01	0.09	0.14	0.05	0.04	0.001	0.1	—	1.0	—	—	2.8	—
440			100.0	0.95	0.92	0.10	0.76	1.12	0.38	0.32	0.005	0.8	—	10.0	—	—	23.0	—
	Brassica spp. Rape																	
441	seeds, meal solvent extracted	5-03-871	91.0	0.61	0.10	0.55	0.95	1.24	0.09	1.14	—	—	—	—	—	1.0	—	—
442			100.0	0.67	0.11	0.60	1.04	1.36	0.10	1.25	—	—	—	—	—	1.1	—	—
	Brevoortia tyrannus. Fish, menhaden																	
443	meal mechanical extracted	5-02-009	92.0	5.18	0.55	0.14	2.89	0.70	0.39	0.45	0.153	10.7	1.09	480.0	33.7	2.2	148.3	—
444			100.0	5.65	0.60	0.16	3.16	0.76	0.43	0.49	0.167	11.7	1.19	524.0	36.8	2.4	161.9	—
	Carthamus tinctorius. Safflower																	
445	seeds without hulls, meal solvent extracted	5-07-959	92.0	0.35	0.16	1.02	1.29	1.10	0.04	0.20	1.969	8.5	—	484.0	39.3	—	33.3	—
446			100.0	0.38	0.18	1.11	1.40	1.19	0.05	0.22	2.147	9.3	—	528.0	42.8	—	36.3	—
	Cereals																	
447	brewers grains, dehydrated	5-02-141	92.0	0.30	0.15	0.15	0.51	0.08	0.21	0.30	0.076	21.2	0.07	245.0	37.2	0.7	27.3	0.5
448			100.0	0.33	0.17	0.16	0.55	0.09	0.23	0.32	0.082	23.0	0.07	266.0	40.4	0.8	29.6	0.5
	Cicer arietinum. Chickpea																	
449	seeds	5-01-218	91.0	0.15	—	—	0.34	0.81	0.03	—	—	—	—	70.0	—	—	—	—
450			100.0	0.17	—	—	0.37	0.89	0.03	—	—	—	—	77.0	—	—	—	—
	Cocos nucifera. Coconut																	
451	meats oil residue, meal solvent extracted	5-01-573	91.0	0.17	0.03	0.33	0.60	1.49	0.04	0.34	0.126	9.4	—	683.0	65.4	—	—	—
452			100.0	0.19	0.03	0.36	0.66	1.63	0.04	0.37	0.139	10.4	—	750.0	71.8	—	—	—
	Cyamopsis tetragonoloba. guar																	
453	seeds without endosperm, ground	5-05-687	88.0	0.19	—	—	0.49	1.46	—	0.39	—	—	—	—	—	—	—	—
454			100.0	0.22	—	—	0.56	1.67	—	0.44	—	—	—	—	—	—	—	—
	Glycine max. Soybean																	
455	seeds, meal solvent extracted	5-04-604	90.0	0.30	0.04	0.27	0.63	1.97	0.34	0.43	0.090	22.8	0.13	119.0	29.0	0.3	42.9	0.2
456			100.0	0.34	0.04	0.30	0.70	2.20	0.38	0.47	0.101	25.4	0.15	133.0	32.4	0.3	47.9	0.2
	Gossypium spp. Cotton																	
457	seeds, meal solvent extracted, 41% protein	5-01-621	91.0	0.17	0.05	0.54	1.10	1.39	0.04	0.26	0.150	19.9	—	208.0	20.8	—	62.2	—
458			100.0	0.18	0.05	0.59	1.21	1.52	0.05	0.28	0.165	21.8	—	228.0	22.8	—	68.2	—

TABLE 3 Composition of Goat Feeds—*Continued*

Entry Number	SCIENTIFIC NAME Feed Name	International Feed Number	Dry Matter (%)	Calcium (%)	Chlorine (%)	Magnesium (%)	Phosphorus (%)	Potassium (%)	Sodium (%)	Sulfur (%)	Cobalt (mg/kg)	Copper (mg/kg)	Iodine (mg/kg)	Iron (mg/kg)	Manganese (mg/kg)	Selenium (mg/kg)	Zinc (mg/kg)	Carotene (mg/kg)
	Helianthus annuus. Sunflower, common																	
459	seeds without hulls,	5-04-739	93.0	0.41	0.10	0.71	0.91	1.06	0.22	—	—	3.5	—	31.0	19.0	—	—	—
460	meal solvent extracted		100.0	0.44	0.11	0.77	0.98	1.14	0.24	—	—	3.8	—	33.0	20.4	—	—	—
	Linum usitatissimum. Flax, common																	
461	seeds, meal solvent	5-02-048	90.0	0.39	0.04	0.60	0.80	1.38	0.14	0.39	0.188	25.7	—	319.0	37.6	0.8	—	—
462	extracted		100.0	0.43	0.04	0.66	0.89	1.53	0.15	0.43	0.208	28.5	—	354.0	41.7	0.9	—	—
	Lycopersicon esculentum. Tomato																	
463	pomace, dehydrated	5-05-041	92.0	0.39	—	0.18	0.55	3.33	—	—	—	29.9	—	4223.0	47.0	—	—	—
464			100.0	0.43	—	0.20	0.60	3.63	—	—	—	32.6	—	4600.0	51.2	—	—	—
	Meat, with bone, meal																	
465	rendered	5-00-388	93.0	10.30	0.74	1.02	5.10	1.33	0.72	0.25	0.180	1.5	1.31	684.0	13.3	0.3	89.1	—
466			100.0	11.06	0.80	1.09	5.48	1.43	0.77	0.27	0.193	1.6	1.41	735.0	14.3	0.3	95.6	—
	Phaseolus vulgaris. Bean, navy																	
467	seeds	5-00-623	89.0	0.16	0.06	0.13	0.52	1.31	0.04	0.23	—	9.9	—	99.0	21.1	—	—	—
468			100.0	0.18	0.06	0.15	0.59	1.47	0.05	0.26	—	11.0	—	110.0	23.6	—	—	—

TABLE 3 Composition of Goat Feeds—*Continued*

Entry Number	SCIENTIFIC NAME Feed Name	International Feed Number	Dry Matter (%)	Calcium (%)	Chlorine (%)	Magnesium (%)	Phosphorus (%)	Potassium (%)	Sodium (%)	Sulfur (%)	Cobalt (mg/kg)	Copper (mg/kg)	Iodine (mg/kg)	Iron (mg/kg)	Manganese (mg/kg)	Selenium (mg/kg)	Zinc (mg/kg)	Carotene (mg/kg)
	Poultry																	
469	feathers, meal	5-03-795	91.0	0.25	0.28	0.20	0.66	0.28	0.69	1.47	0.043	6.4	0.04	74.0	12.5	0.8	67.8	—
470	hydrolyzed		100.0	0.28	0.30	0.22	0.72	0.31	0.76	1.61	0.047	7.0	0.05	81.0	13.6	0.9	74.3	—
	Secale cereale. Rye																	
471	distillers grains,	5-04-023	92.0	0.15	0.05	0.17	0.48	0.07	0.17	0.44	—	—	—	—	18.4	—	—	—
472	dehydrated		100.0	0.16	0.05	0.18	0.52	0.08	0.18	0.48	—	—	—	—	20.0	—	—	—
	Sesamum indicum. Sesame																	
473	seeds, meal mechanical	5-04-220	93.0	2.01	0.07	0.46	1.36	1.25	0.04	0.33	—	—	—	93.0	47.8	—	99.7	0.4
474	extracted		100.0	2.17	0.07	0.50	1.46	1.35	0.04	0.35	—	—	—	100.0	51.5	—	107.5	0.5
	Sorghum bicolor. Sorghum																	
475	distillers grains,	5-04-374	94.0	0.15	—	0.18	0.69	0.36	0.05	0.17	—	—	—	47.0	—	—	—	—
476	dehydrated		100.0	0.16	—	0.19	0.74	0.38	0.05	0.18	—	—	—	50.0	—	—	—	—
	Triticum aestivum. Wheat																	
477	germs, ground	5-05-218	88.0	0.05	0.07	0.25	0.92	0.97	0.03	0.25	0.118	9.7	—	51.0	133.7	0.4	119.0	—
478			100.0	0.06	0.08	0.28	1.05	1.09	0.03	0.28	0.134	10.9	—	58.0	151.4	0.4	134.8	—

TABLE 4 Composition Values of Minerals for Goats; Data Expressed As-Fed and Dry (100% Dry Matter)[a]

Entry Number	Feed Name Description	International Feed Number	Dry Matter (%)	Protein Equivalent $N \times 6.25$ (%)	Calcium (Ca) (%)	Chlorine (Cl) (%)	Magnesium (Mg) (%)	Phosphorus (P) (%)	Potassium (K) (%)
	AMMONIUM								
001	phosphate, monobasic,	6-09-338	97	68.8	0.27	—	0.45	24.00	0.01
002	$(NH_4)H_2PO_4$		100	70.9	0.28	—	0.46	24.74	0.01
003	phosphate, dibasic,	6-00-370	97	112.4	0.50	—	0.45	20.00	0.01
004	$(NH_4)_2HPO_4$		100	115.9	0.52	—	0.46	20.60	0.01
	BONE								
005	charcoal (Bone black)	6-00-402	90	8.5	27.10	—	0.53	12.73	0.14
006	(Bone char)		100	9.4	30.11	—	0.59	14.14	0.16
007	meal, steamed	6-00-400	97	12.8	29.82	—	0.32	12.49	0.18
008			100	13.2	30.71	—	0.33	12.86	0.19
	CALCIUM								
009	carbonate, $CaCO_3$	6-01-069	100	—	39.39	—	0.05	0.04	0.06
010			100	—	39.39	—	0.05	0.04	0.06
011	phosphate, monobasic, from	6-01-082	97	—	15.91	—	0.59	20.95	0.08
012	defluorinated phosphoric acid		100	—	16.40	—	0.61	21.60	0.08
013	phosphate, dibasic, from	6-01-080	97	—	21.30	—	0.57	18.70	0.07
014	defluorinated phosphoric acid		100	—	22.00	—	0.59	19.30	0.07
015	sulfate, anhydrous (Gypsum),	6-01-087	85	—	22.02	—	2.21	0.01	—
016	$CaSO_4$		100	—	25.90	—	2.61	0.01	—
	COBALT								
017	carbonate, $CoCO_3$	6-01-566	99[b]	—	—	—	—	—	—
018			100	—	—	—	—	—	—
019	COLLOIDAY CLAY (Soft	6-03-947	99[b]	—	17.00	—	0.38	9.00	—
020	rock phosphate)		100	—	17.17	—	0.38	9.09	—
	COPPER (CUPRIC)								
021	sulfate, pentahydrate	6-01-720	100	—	—	—	—	—	—
022	$CuSO_4 \cdot 5H_2O$, cp		100	—	—	—	—	—	—
023	CURACAO PHOSPHATE	6-05-586	99[b]	—	34.00	—	0.80	14.00	—
024			100	—	34.34	—	0.81	14.14	—
	IRON (FERROUS)								
025	sulfate, heptahydrate,	6-20-734	98[b]	—	—	—	—	—	—
026	$FeSO_4 \cdot 7H_2O$		100	—	—	—	—	—	—
	LIMESTONE								
027	limestone	6-02-632	100	—	34.00	0.03	2.06	0.02	0.12
028			100	—	34.00	0.03	2.06	0.02	0.12
029	magnesium (dolomitic)	6-02-633	99[b]	—	22.08	0.12	9.89	0.04	0.36
030			100	—	22.30	0.12	9.99	0.04	0.36
	MAGNESIUM								
031	carbonate, $MgCO_3Mg(OH)_2$	6-02-754	98[b]	—	0.02	0.00	30.20	—	—
032			100	—	0.02	0.00	30.81	—	—
033	oxide, MgO	6-02-756	98	—	3.00	—	54.90	—	—
034			100	—	3.07	—	56.20	—	—

Entry Number	Sodium (Na) (%)	Sulfur (S) (%)	Cobalt (Co) (%)	Copper (Cu) (%)	Fluorine (F) (%)	Iodine (I) (%)	Iron (Fe) (%)	Manganese (Mn) (%)	Selenium (Se) (%)	Zinc (Zn) (%)
001	0.06	1.42	0.001	0.001	0.24	—	1.69	0.04	—	0.01
002	0.06	1.46	0.001	0.001	0.25	—	1.74	0.04	—	0.01
003	0.05	2.10	—	0.001	0.20	—	1.20	0.04	—	0.01
004	0.05	2.16	—	0.001	0.21	—	1.24	0.04	—	0.01
005	—	—	—	—	—	—	—	—	—	—
006	—	—	—	—	—	—	—	—	—	—
007	5.53	2.44	—	—	—	—	2.600	—	—	0.01
008	5.69	2.51	—	—	—	—	2.670	—	—	0.01
009	0.06	—	—	—	—	—	0.030	0.03	—	—
010	0.06	—	—	—	—	—	0.030	0.03	—	—
011	0.06	1.19	0.001	0.001	0.20	—	1.53	0.035	—	0.009
012	0.06	1.22	0.001	0.001	0.21	—	1.58	0.036	—	0.009
013	0.05	1.11	0.001	0.001	0.18	—	1.40	0.03	—	0.01
014	0.05	1.14	0.001	0.001	0.18	—	1.44	0.03	—	0.01
015	—	20.01	—	—	—	—	0.171	—	—	—
016	—	23.54	—	—	—	—	0.201	—	—	—
017	—	0.20	45.54	—	—	—	0.049	—	—	—
018	—	0.20	46.00	—	—	—	0.050	—	—	—
019	0.10	—	—	—	1.49	—	1.90	—	—	—
020	0.10	—	—	—	1.50	—	1.92	—	—	—
021	—	12.84	—	25.45	—	—	—	—	—	—
022	—	12.84	—	25.45	—	—	—	—	—	—
023	0.20	—	—	—	0.54	—	0.35	—	—	—
024	0.20	—	—	—	0.55	—	0.35	—	—	—
025	—	12.10	—	—	—	—	21.400	—	—	—
026	—	12.35	—	—	—	—	21.840	—	—	—
027	0.06	0.04	—	—	—	—	0.350	—	—	—
028	0.06	0.04	—	—	—	—	0.350	—	—	—
029	—	—	—	—	—	—	0.076	—	—	—
030	—	—	—	—	—	—	0.077	—	—	—
031	—	—	—	—	—	—	0.021	—	—	—
032	—	—	—	—	—	—	0.022	—	—	—
033	—	—	—	—	0.02	—	—	0.01	—	—
034	—	—	—	—	0.02	—	—	0.01	—	—

TABLE 4 Composition Values of Minerals for Goats; Data Expressed As-Fed and Dry (100% Dry Matter)[a]—*Continued*

Entry Number	Feed Name Description	International Feed Number	Dry Matter (%)	Protein Equivalent $N \times 6.25$ (%)	Calcium (Ca) (%)	Chlorine (Cl) (%)	Magnesium (Mg) (%)	Phosphorus (P) (%)	Potassium (K) (%)
	MANGANESE (MANGANOUS)								
035	oxide, MnO, cp	6-03-056	99[b]	—	—	—	—	—	—
036			100	—	—	—	—	—	—
037	OYSTERSHELL, GROUND	6-03-481	99	—	37.62	0.01	0.30	0.07	0.10
038	(flour)		100	—	38.00	0.01	0.30	0.07	0.10
	PHOSPHATE								
039	rock, defluorinated	6-01-780	100	—	32.00	—	0.42	18.00	0.09
040			100	—	32.00	—	0.42	18.00	0.09
	POTASSIUM								
041	bicarbonate, $KHCO_3$, cp	6-29-493	99[b]	—	—	—	—	—	38.65
042			100	—	—	—	—	—	39.05
043	iodide, KI	6-03-759	100[b]	—	—	—	—	—	21.00
044			100	—	—	—	—	—	21.00
	SODIUM								
045	bicarbonate, $NaHCO_3$	6-04-272	100	—	—	—	—	—	—
046			100	—	—	—	—	—	—
047	chloride, NaCl	6-04-152	100	—	—	60.66	—	—	—
048			100	—	—	60.66	—	—	—
049	phosphate, monobasic,	6-04-288	97	—	—	—	—	21.80	—
050	monohydrate, $NaH_2PO_4 \cdot H_2O$		100	—	—	—	—	22.50	—
051	selenite, Na_2SeO_3	6-26-013	98[b]	—	—	—	—	—	—
052			100	—	—	—	—	—	—
053	sulfate, decahydrate,	6-04-292	97[b]	—	—	—	—	—	—
054	$Na_2SO_4 \cdot 10H_2O$, cp		100	—	—	—	—	—	—
055	tripolyphosphate, $Na_5P_3O_{10}$	6-08-076	96	—	—	—	—	24.00	—
056			100	—	—	—	—	25.00	—
	ZINC								
057	oxide, ZnO	6-05-553	100	—	—	—	—	—	—
058			100	—	—	—	—	—	—
059	sulfate, monohydrate,	6-05-555	99[b]	—	0.02	0.015	—	—	—
060	$ZnSO_4 \cdot H_2O$		100	—	0.02	0.015	—	—	—

[a] The composition of mineral ingredients that are hydrated (e.g. $CaSO_4 \cdot 2H_2O$) is shown including the waters of hydration, both on an as-fed and dry matter basis. Mineral composition of feed grade mineral supplements varies by source, mining site, and manufacturer. Use manufacturer's analysis when available.
[b] Dry matter values have been estimated for these minerals.

Entry Number	Sodium (Na) (%)	Sulfur (S) (%)	Cobalt (Co) (%)	Copper (Cu) (%)	Fluorine (F) (%)	Iodine (I) (%)	Iron (Fe) (%)	Manganese (Mn) (%)	Selenium (Se) (%)	Zinc (Zn) (%)
035	—	—	—	—	—	—	—	76.67	—	—
036	—	—	—	—	—	—	—	77.45	—	—
037	0.21	—	—	—	—	—	0.284	0.01	—	—
038	0.21	—	—	—	—	—	0.287	0.01	—	—
039	0.04	—	0.001	0.002	0.18	—	0.670	0.02	—	0.006
040	0.04	—	0.001	0.002	0.18	—	0.670	0.02	—	0.006
041	—	—	—	—	—	—	—	—	—	—
042	—	—	—	—	—	—	—	—	—	—
043	—	—	—	—	—	68.17	—	—	—	—
044	—	—	—	—	—	68.17	—	—	—	—
045	27.00	—	—	—	—	—	—	—	—	—
046	27.00	—	—	—	—	—	—	—	—	—
047	39.34	—	—	—	—	—	—	—	—	—
048	39.34	—	—	—	—	—	—	—	—	—
049	16.18	—	—	—	—	—	—	—	—	—
050	16.18	—	—	—	—	—	—	—	—	—
051	26.07	—	—	—	—	—	—	—	44.7	—
052	26.60	—	—	—	—	—	—	—	45.6	—
053	13.84	9.65	—	—	—	—	—	—	—	—
054	14.27	9.95	—	—	—	—	—	—	—	—
055	29.80	—	—	—	—	—	0.004	—	—	—
056	31.00	—	—	—	—	—	—	—	—	—
057	—	—	—	—	—	—	—	—	—	78.00
058	—	—	—	—	—	—	—	—	—	78.00
059	—	17.50	—	—	—	—	0.001	0.001	—	36.00
060	—	17.68	—	—	—	—	0.001	0.001	—	36.36

TABLE 5 List of Scientific and Common Names

Dry Forages and Roughages

Scientific name	Common name
Agropyron desertorum	Wheatgrass, crested
Agrostis alba	Redtop
Andropogon spp	Bluestem
Arachis hypogaea	Peanut, groundnut
Avena sativa	Oats
Brachiaria mutica	Paragrass
Bromus inermis	Brome, smooth
Cajanus cajan	Pigeonpea
Carthamus tinctorius	Safflower
Cicer arietinum	Chickpea, gram
Chloris gayana	Rhodesgrass
Cynodon dactylon	Bermudagrass
Cynodon dactylon	Bermudagrass, coastal
Dactylis glomerata	Orchardgrass, cocksfoot
Digitaria decumbens	Pangolagrass
Echinochloa crusgalli	Barnyardgrass
Eragrostis spp	Lovegrass
Festuca arundinacea	Fescue, Kentucky 31
Glycine max	Soybean
Gossypium spp	Cotton
Hordeum vulgare	Barley
Kochia spp	Summercypress
Lespedeza striata	Lespedeza, common
Lolium multiflorum	Ryegrass, Italian
Medicago sativa	Alfalfa, Lucerne
Oryza sativa	Rice
Panicum maximum	Guineagrass
Paspalum dilatatum	Dallisgrass
Paspalum notatum	Bahiagrass
Pennisetum purpureum	Napiergrass
Phalaris arundinacea	Canarygrass, reed
Phleum pratense	Timothy
Pisum sativum sativum	Pea, garden
Poa pratensis	Bluegrass, Kentucky
Prosopis spicigera	Mesquite, spicigera
Pueraria spp	Kudzu
Saccharum officinarum	Sugarcane
Secale cereale	Rye
Sorghum bicolor	Sorghum
Sorghum bicolor sudanense	Sorghum, Sudangrass
Sorghum halepense	Sorghum, Johnsongrass
Trifolium alexandrinum	Clover, Egyptian
Trifolium pratense	Clover, red
Triticum aestivum	Wheat
Vicia spp	Vetch
Vigna sinensis	Cowpea, common
Wood	
Zea mays	Corn, maize
Ziziphus nummularia	Jujube, Wild

Pasture, Range Plants, and Forages Fed Green

Scientific name	Common name
Acacia catechu	Acacia, catechu
Agave spp	Agave
Agropyron desertorum	Wheatgrass, crested
Agropyron smithii	Wheatgrass, western
Agrostis alba	Redtop
Ailanthus excelsa	Ailanthus, excelsa
Albizia lebbeck	Lebbeck tree
Ambrosia spp	Ragweed
Andropogon barbinodis	Bluestem, cane
Andropogon scoparius	Bluestem, little
Anemone heterophylla	Anemone, heterophylla
Aristida oligantha	Threeawn, prairie
Aristida wrightii	Threeawn, Wright
Artemisia filifolia	Sagebrush, sand
Artemisia ludoviciana albula	Sagebrush, Mexican
Aster spp	Aster
Astragalus nuttallii	Milkvetch, Nuttall
Atriplex confertifolia	Saltbush, Shadscale
Avena sativa	Oats
Azadirachta indica	Margosa
Bouteloua curtipendula	Grama, sideoats
Bouteloua hirsuta	Grama, hairy
Brachiaria mutica	Paragrass
Brassica campestris	Rape, turnip
Brassica oleracea medullosa	Kale, marrow
Bromus catharticus	Brome, rescue
Buchloe dactyloides	Buffalograss
Carex planostachys	Sedge, thicket
Carthamus tinctorius	Safflower
Celtis reticulata	Hackberry, netleaf
Celtis tetendra	Hackberry, tetendra
Cenchrus ciliaris	Buffelgrass
Cenchrus spp	Sandbur
Chloris spp	Windmillgrass
Cichorium intybus	Chicory, bluedaisy
Commelina spp	Dayflower
Crotalaria juncea	Crotalaria, sunn
Corton spp	Croton
Cucurbita foetidissima	Buffalogourd
Cyamopsis tetragonoloba	Guar
Cynodon dactylon	Bermudagrass
Dactylis glomerata	Orchardgrass, cocksfoot
Dactyloctenium aegyptium	Crowfootgrass, Durban
Dalbergia sissoo	Dalbergia, sissoo
Dalea formosa	Dalea, feather
Dasylirion texanum	Sotol, Texas
Digera aneensis	Digera, aneensis
Diospyros texana	Persimmon, Texas
Echinochloa crusgalli	Barnyardgrass
Elymus canadensis	Wildrye, Canada
Elymus junceus	Wildrye, Russian
Engelmannia pinnatifida	Engelmanndaisy
Eriochloa sericea	Cupgrass, Texas
Eurotia lanata	Winterfat, common

Evax prolifera	Evax, prolifera
Ficus benghalensis	Banyantree
Ficus glomerata	Fig, cluster
Ficus religiosa	Peepultree
Ficus roxburghii	Fig, roxburgh
Ficus virens	Fig, spotted
Forestiera pubescens	Forestiera, downy
Grewia oppositifolia roxburghii	Grewia, oppositifolia, roxburghii
Gutierrezia texana	Snakeweed, Texas
Gymnosperia spinosa	Gymnosperia, spinosa
Helianthus spp	Sunflower
Hilaria belangeri	Curlymesquite
Hordeum vulgare	Barley
Juniperus ashei	Juniper, ashe
Juniperus pinchotii	Juniper, redberry
Kochia vestita	Summercypress, gray
Lactuca sativa	Lettuce
Lasiurus sindicus	Lasiurus, sindicus
Lepidium spp	Pepperweed
Leptochloa dubia	Sprangletop, green
Leptoloma cognatum	Witchgrass, fall
Lespedeza striata	Lespedeza, common
Lesquerella gordonii	Bladderpod, Gordon
Leucaena retusa	Leadtree, littleleaf
Lolium spp	Ryegrass
Lonicera albiflora	Honeysuckle, bush
Lotus spp	Trefoil
Lupinus texensis	Bluebonnet, Texas
Medicago sativa	Alfalfa, Lucerne
Melia azedarach	Chinaberry
Mimosa spp	Mimosa
Moringa oleifera	Horseradishtree
Morus australis	Mulberry, Japanese
Morus spp	Mulberry
Musa spp	Banana
Nolina texana	Nolina, Texas
Onobrychis viciifolia	Sainfoin, common
Opuntia spp	Pricklypear
Panicum antidotale	Panicum, blue
Panicum hallii	Panicum, Halls
Panicum maximum	Guineagrass
Panicum obtusum	Vine-mesquite
Panicum virgatum	Switchgrass
Paspalum dilatatum	Dallisgrass
Pennisetum incomptum	Pennisetum, incomptum
Pennisetum purpureum	Napiergrass
Phalaris tuberosa stenoptera	Hardinggrass
Phaseolus atropurpureus	Bean, purple
Pisum sativum sativum	Pea, garden
Pisum spp	Pea
Plantago rhodosperma	Plantain, redseed
Prosopis glandulosa	Mesquite, honey
Quercus sinuata breviloba	Oak, shin
Quercus virginiana	Oak, live
Ratibida columnifera	Prairieconeflower, upright
Rhus aromatica	Sumac, fragrant
Sarcobatus vermiculatus	Greasewood, black
Schrankia roemeriana	Sensitivebrier, Roemer
Secale cereale	Rye
Sedum nuttallianum	Stonecrop, Nuttall
Sesbania cannabina	Sesbania, pea
Setaria macrostachya	Bristlegrass, plains
Setaria spp	Millet
Sorghum bicolor	Sorghum
Sorghum bicolor sudanense	Sorghum, Sudangrass
Sorghum halepense	Sorghum, Johnsongrass
Stipa leucotricha	Needlegrass, Texas
Syzygium cumini	Javaplum
Tillandsia usneoides	Spanishmoss
Tribulus terrestris	Puncturevine
Trifolium alexandrinum	Clover, Egyptian, Alexandrian, Berseem
Triodia albescens	Triodia, white
Vicia sativa	Vetch, common
Vigna sinensis	Cowpea, common
Yucca spp	Yucca
Zexmenia hispida	Zexmenia, orange
Ziziphus jujuba	Jujube, common

Silages

Avena sativa	Oats
Beta vulgaris altissima	Beet, sugar
Digitaria decumbens	Pangolagrass
Glycine max	Soybean
Medicago sativa	Alfalfa
Pennisetum purpureum	Napiergrass
Pisum spp	Pea
Sorghum bicolor saccharatum	Sorghum, sorgo
Zea mays	Corn, maize

Energy Feeds

Animal

Avena sativa	Oats
Beta vulgaris altissima	Beet, sugar
Bos taurus	Cattle
Brassica oleracea capitata	Cabbage
Brassica rapa rapa	Turnip
Ceratonia siliqua	Carob bean
Citrus spp	Citrus
Daucus spp	Carrot
Hordeum vulgare	Barley
Ipomoea batatas	Sweetpotato
Laminariales (order)—Fucales (order)	Seaweed, kelp
Malus spp	Apple
Manihot esculenta	Cassava, common
Oryza sativa	Rice
Pennisetum glaucum	Pearlmillet
Quercus spp	Oak
Saccharum officinarum	Sugarcane
Secale cereale	Rye
Solanum tuberosum	Potato

TABLE 5 List of Scientific and Common Names—*Continued*

Sorghum bicolor subglabrescens	Sorghum, milo	*Cyamopsis tetragonoloba*	Guar
Triticum aestivum	Wheat		
Zea mays	Corn, maize	*Glycine max*	Soybean
Zea mays indentata	Corn, dent yellow	*Gossypium* spp	Cotton
		Helianthus annuus	Sunflower, common
Protein Supplements		*Linum usitatissimum*	Flax, common
		Lycopersicon esculentum	Tomato
Arachis hypogaea	Peanut, groundnut	Meat, With Bone, Meal Rendered	
Blood, Meal Spray Dehy (Blood Flour)			
		Phaseolus vulgaris	Bean, Navy
Bos taurus	Cattle	Poultry	
Brassica spp	Rape	*Secale cereale*	Rye
Brevoortia tyrannus	Fish, menhaden	*Sesamum indicum*	Sesame
Carthamus tinctorius	Safflower	*Sorghum bicolor*	Sorghum
Cereals		*Triticum aestivum*	Wheat
Cicer arietinum	Chickpea, gram	Urea	
Cocos nucifera	Coconut		

BIBLIOGRAPHY

INTRODUCTION

Ace, D. L. 1978. Dairy goats require lots of care just to break even, p. 357. *In* Yearbook of Agriculture. U.S. Department of Agriculture, Washington, D.C.

Agraz, Garcia, A. A. 1958. La cabra lechera, fuente de riqueza para el campesino. (The dairy goat, a source of wealth for the small farmer.) Bartolome Trucco Publ., Mexico City, Mexico.

Agricultural Research Institute Cyprus. 1964. Cost and returns of sheep and goats. Farm Manage. Rep. 3.

Anonymous. 1973. Indagine nazionale su alcuni aspetti degli allevamenti e delle produzioni caprine. (National research on aspects of dairy goat production.) Ministero dell' Agricoltura e delle Foreste, Associazione Nazionale della Pastorizia, Rome, Italy.

Banco Nacional Agropecuario. 1971. Goat husbandry: An important animal resource. Mexico City, Mexico.

Benzie, D., and A. T. Phillipson. 1957. The alimentary tract of the ruminant. Oliver and Boyd, Edinburgh, Scotland.

Bhattacharya, A. N. 1980. Research on goat nutrition and management in Mediterranean Middle East, and adjacent Arab countries. J. Dairy Sci. 63:1681.

Brandsch, H., and W. Kruger. 1970. Stand und zuechterische Entwicklungsmoeglicheiten der Ziegen- und Milchschafzucht. (Situation and possibilities of development in goat and milk sheep breeding.) Arch. Gefluegelzucht Kleintierkunde. 19:255.

Bula, R. J., V. L. Lechtenberg, D. A. Holt, L. R. Humphreys, L. V. Crowder, and T. W. Box. 1977. Potential of the world's forages for ruminant animal production. Winrock Int. Livestock Res. Training Center, Morrilton, Ark.

Choveiri, G. 1973. The Damascus goat, the future milking animal of the Mediterranean countries. *In* Proc. 3rd World Conf. Anim. Prod., vol. 1, Melbourne, Australia.

Clamohoy, L. L., V. G. Arganosa, and O. A. Palad. 1959. Exploratory goat dairying with native and grade goats. Philipp. Agric. 43:243.

Corbett, J. A. 1978. The goats' role in subsistence development. Dairy Goat J. 56:7.

Crawford, L. V., and F. T. Grogan. 1961. Allergenicity of cow's milk proteins. IV. Relationship to goats' milk proteins as studied by serum-agar precipitation. J. Pediatr. 59:347.

Devendra, C. 1975. Sheep and goat production in Guyana. Z. Tierz. Zuechtungsbiol. 92:305.

Devendra, C. 1979. Goat and sheep production potential in the ASEAN region. World Anim. Rev. 32:33.

Devendra, C. 1980. Milk production in goats compared to buffalo and cattle in humid tropics. J. Dairy Sci. 63:1755.

Devendra, C., and M. Burns. 1970. Goat production in the tropics. Commonw. Bur. Anim. Breed. Genet. Tech. Commun. 19.

Devendra, C., and K. Nozawa. 1976. Goats in South East Asia—their status and production. Z. Tierz. Zuechtungsbiol. 93:101.

Eker, M. 1961. Problems arising from nomadism and transhumance of goat flocks in Turkey, p. 120. *In* Yearbook faculty agriculture. University of Ankara, Turkey.

Fitzhugh, H. A., H. J. Hodgson, O. J. Scoville, T. D. Nguyen, and T. C. Byerly. 1978. The role of ruminants in support of man. Winrock Int. Livestock Res. Training Center, Morrilton, Ark.

Food and Agriculture Organization. 1979. Production yearbook. FAO, Rome, Italy.

French, M. H. 1970. Observations on the goat. Food Agric. Org. Agric. Stud. 80, Rome, Italy.

Gall, C. 1980. Relationship between body conformation and production in dairy goats. J. Dairy Sci. 63:1768.

Gamble, J. A., N. R. Ellis, and A. K. Besley. 1939. Composition and properties of goat's milk as compared with cow's milk. USDA Techn. Bull. 671.

Grieder, H. R., and A. Buergin-Wolff. 1961. Development of allergy towards several nutrient materials in the same infant with detection of the corresponding antibodies. Schweiz. Med. Wochenschr. 91:747.

Guss, S. 1977. Management and diseases of dairy goats. Dairy Goat J. Publ., Scottsdale, Ariz.

Haenlein, G. F. W. 1978. Dairy goat management. J. Dairy Sci. 61:1011.

Haenlein, G. F. W. 1980a. Status of world literature on dairy goats—introductory remarks. J. Dairy Sci. 63:1591.

Haenlein, G. F. W. 1980b. Nutrient requirements of goats, past and present. Int. Goat Sheep Res. 1:79.

Haenlein, G. F. W. 1981. Dairy goat industry of the United States. J. Dairy Sci. 64:1288.

Hanson, L. A., and H. J. Andersen. 1962. A comparison of the antigenic relationships of human milk and goats' milk to bovine milk. Acta Paediatr. Stockholm 51:509.

Harris, E. 1980. Southern agriculture corporations experience in marketing goat dairy products. J. Dairy Sci. 63:1649.

Hofmeyr, H. S., D. M. Joubert, F. J. G. Badenhorst, and G. J. van D. Steyn. 1965. Adaptability of sheep and goats to a South African tropical environment. Proc. S. Afr. Soc. Anim. Prod. 4:191.

Iloeje, M. U., T. R. Rounsaville, R. E. McDowell, G. R. Wiggans, and L. D. Van Vleck. 1980. Age-season adjustment factors for Alpine, LaMancha, Nubian, Saanen and Toggenburg dairy goats. J. Dairy Sci. 63:1309.

Institut Technique de l'Elevage Ovin et Caprin. 1971. Proc. 2nd Internat. Conf. Goat Breeding at Tours. Inst. Techn. de l'Elevage Ovin Caprin Publ., Paris, France.

Institut Technique de l'Elevage Orin et Caprin. 1981. Proc. Symp. Internat. Nutrition and Systems of Goat Feeding at Tours. Inst. Techn. de l'Elevage Orin Caprin Publ., Paris, France.

Jenness, R. 1980. Composition and characteristics of goat milk: Review 1968–1979. J. Dairy Sci. 63:1605.

Joubert, D. M. 1973. Goats in the animal agriculture of Southern Africa. Z. Tierz. Zuechtungsbiol. 90:245.

Larson, B. L. 1978. The dairy goat as a model in lactation studies. J. Dairy Sci. 61:1023.

Leach, K. 1980. Milk production trends in dairy goats. J. Dairy Sci. 63:1600.

Loewenstein, M., S. J. Speck, H. M. Barnhart, and J. F. Frank. 1980. Research on goat milk products: A review. J. Dairy Sci. 63:1631.

Lowe., H. J. 1943. Goats as a source of milk in tsetse areas. E. Afr. Agric. J. 8:238.

Mack, P. B. 1953. A preliminary nutrition study of the value of goats' milk in the diet of children. Am. Goat Soc. Yearb.

Maubecin, R. A. 1973. La explotacion del ganado caprino en la Republica Argentina. (Exploitation of goat breeding in the Argentine Republic.) Inst. Nac. Tecnol. Agropecu. 55:25.

Maule, J. P. 1966. A note on dairy goats in the tropics. Anim. Breed. Abstr. 34:153.

McDowell, R. E., and L. Bove. 1977. The goat as a producer of meat. Cornell Int. Agric. Mimeo 56.

Meklenburcev, R. N. 1949. The ecology of the Siberian mountain goat in the Pamirs. Zool. Z. 28:482.

Morand-Fehr, P., and M. deSimiane. 1977. Proc. Nutr. Goats Mediterran., Granada, Spain.

National Research Council, Institute of Laboratory Animal Resources. 1974. Ruminants, cattle, sheep and goats: Guidelines for the breeding, care and management of laboratory animals. National Academy of Sciences, Washington, D.C.

Nottbohm, F. E., and K. Phillipi. 1933. Die Sonderstellung der Ziegenmilch und die Zusammensetzung des Ziegenblutes. (The unique position of goat milk and the composition of goat blood.) Z. Unters. Lebensm. 66:289.

Oppong, E. N. W. 1965. A note on goats in Ghana with reference to the need to develop goat husbandry to improve the nation's diet. Ghana Fmr. 9:144.

Parkash, S., and R. Jenness. 1968. The composition and characteristics of goats' milk: A review. Dairy Sci. Abstr. 30:67.

Rogers, A. L., L. F. Erickson, A. S. Hoversland, Y. Metcalfe, and P. L. Clary. 1969. Management of a colony of African pygmy goats for biomedical research. Yale Anim. Care 19:181.

Sands, M., and R. E. McDowell. 1978. The potential of the goat for milk production in the tropics. Cornell Int. Agric. Mimeo 60.

Sanfiorenzo, J. 1962. Evaluation of six breeding lines of milk goats. J. Agric. Univ. P.R. 46:205.

Saperstein, S. 1960. Antigenicity of the whey proteins in evaporated cow's milk and whole goat's milk. Ann. Allergy 18:765.

Schaefer, H. 1967. Sheep and goats as rivals in the arid savannah. Proc. Techn. Econ. Probl. Prod. Sheep, Goats, Fur-bearing Anim. Milan, Italy.

Shannon, J. L. 1956. Care and management of dairy goats in Trinidad and Tobago. Department of Agriculture Bull. 9, Trinidad and Tobago.

Shelton, M. 1980a. Angora goat and mohair production, p. 362. *In* M. E. Ensminger, Sheep and goat handbook, vol. 1. Agriservices Foundation, Clovis, California.

Shelton, M. 1980b. Meat goat production, p. 369. *In* M. E. Ensminger, Sheep and goat handbook, vol. 1. Agriservices Foundation, Clovis, California.

Shkolnik, A., E. Maltz, and S. Gordin. 1980. Desert conditions and goat milk production. J. Dairy Sci. 63:1749.

Singh, S. N., and O. P. S. Sengar. 1970. Investigation on milk and meat potentialities of Indian goats, 1965–70. Final Techn. Report Project A7-AH-18. Raja Balwant Singh College, Bichpuri (Agra), India.

Skinner, J. D. 1972. Utilisation of the Boer goat for intensive animal production. Trop. Anim. Health Prod. 4:120.

Terrill, C. E. 1980. Future of sheep and goat production in the United States, p. 88. *In* M. E. Ensminger, Sheep and goat handbook, vol. 1. Agriservices Foundation, Clovis, California.

Wahid, A. 1965. Goat raising in Pakistan. Agric. Pakistan 16:509.

Woinoff, O. 1949. Die Entwicklung der westdeutschen Ziegenzucht in den Jahren 1938–1948 und ihre volkswirtschaftliche Bedeutung. (Development of the West German goat industry 1938–1948 and its economic significance.) Der Ziegenzuechter. 40:149.

Zorn, W., H. Brueggemann, G. Comberg, and A. Ott. 1938. Ziege oder Milchschaf in den Kleinstbetrieben. (Dairy goats or milk sheep in part-time farming.) Z. Tierz. Zuechtungbiol. 41:257.

ENERGY

Adu, I. F. 1975. The effect of steaming up on the birth weight, lactation and growth of West African dwarf sheep. Ph.D. thesis. University of Ibadan, Nigeria.

Agricultural Research Council. 1965. Nutrient requirements of farm livesotck. 2. Ruminant. Agric. Res. Counc., London, England.

Akinsoyinu, A. O. 1974. Studies on protein and energy utilization by the West African dwarf goats. Ph.D. thesis. University of Ibadan, Nigeria.

Akinsoyinu, A. O., A. U. Mba, and F. O. Olubaja. 1978. Studies on energy and protein utilization for pregnancy and lactation by the West African dwarf goats in Nigeria. Dairy Sci. Abstr. 40:113.

Beger, C. 1907. Untersuchungen ueber die Einwirkung von Nahrungsfett als Emulsion und als Substanz auf die Milchproduktion. (Studies of the effects of fat as emulsion or unemulsified on milk production.) Landwirtsch. Vers. Stn. 67:1.

Beger, C., O. Doll, and G. Fingerling. 1901. Fuetterungsversuche mit Milchschafen und Ziegen ueber den Einfluss des Nahrungsfettes auf Menge und Zusammensetzung der Milch. (Feeding studies with milk sheep and goats on the effects of feed fat upon milk yield and composition.) Chem. Z. 31:951.

Bender, R. C., and L. A. Maynard. 1932. Fat metabolism in goats. J. Dairy Sci. 15:242.

Blaxter, K. L. 1967. The energy metabolism of ruminants. Hutchinson, London, England.

Brody, S. 1938. Growth and development with special reference to domestic animals. Vol. 49. Growth, milk production, energy metabolism, and energetic efficiency of milk production in goats. Mo. Agric. Exp. Stn. Res. Bull. 291.

Brody, S. 1945. Bioenergetics and growth. Reinhold, New York.

Crampton, E. W. 1963. Nutrition to caloric ratios. J. Nutr. 82:3.

Cunningham, H. M., and J. K. Loosli. 1954. The effect of fat-free diets on lambs and goats. J. Anim. Sci. 13:265.

Devendra, C. 1967a. Studies on the nutrition of the indigenous goat of Malaysia. I. The maintenance requirements of pen-fed goats. Malays. Agric. J. 46:80.

Devendra, C. 1967b. Studies in the nutrition of the indigenous goat of Malaysia. II. The requirements for live weight gain. Malays. Agric. J. 46:98.

Devendra, C. 1975. Biological efficiency of milk production in dairy goats. World Rev. Anim. Prod. 11:46.

Devendra, C., and M. Burns. 1970. Goat production in the tropics. Commonw. Bur. Anim. Breed. Genet. Tech. Commun. 19.

Evans, R. E. 1960. Rations for livestock. 15th Ed. Minist. Agric. Fish Food Bull. 48, London, England.

Fehr, P. M., and J. Delage. 1973. Effet du niveau energetique du regime sur l'utilisation par la mamelle de chevre de l'acetate comme precurseur des acides gras du lait. (The effect of energy level of diet on the utilization by goat mammary gland of acetate as a precursor of fatty acids in milk.) C. R. Acad. Sci. 267D:3449.

Flatt, W. P., P. W. Moe, L. A. Moore, K. Breirem, and A. Ekern. 1972. Energy requirements of cows for lactation, p. 385. *In* W. Lenkeit and K. Breirem, Handbook of animal nutrition, vol. 2. P. Parey, Hamburg, West Germany.

Gallagher, J. R., and M. Shelton. 1972. Efficiencies of conversion of feed to fiber of Angora goats and Rambouillet sheep. J. Anim. Sci. 34:319.

Garrett, W. N., J. H. Mayer, and G. P. Lofgreen. 1959. The comparative energy requirements of sheep and cattle for maintenance and gain. J. Anim. Sci. 18:228.

Haenlein, G. F. W. 1950. Stoffwechsel und Energiehaushalt der Ziege. (Nutrition and energy metabolism of goats.) Thesis, University of Hohenheim-Stuttgart, West Germany.

Haenlein, G. F. W. 1980. Nutrient requirements of dairy goats past and present. Int. Goat Sheep Res. 1:79.

Huston, J. E. 1978. Symposium: Dairy goats. Forage utilization and nutrient requirements of the goat. J. Dairy Sci. 61:988.

Huston, J. E., M. Shelton, and W. C. Ellis. 1971. Nutritional requirements of the Angora goat. Tex. Agric. Exp. Stn. Bull. 1105.

Hutjens, M. F., and L. H. Schultz. 1971. Effect of feeding soybeans or formaldehyde-treated soybeans on lipid metabolism in ruminants. J. Dairy Sci. 54:1876.

Itoh, M., T. Haryu, R. Tano, and K. Iwasaki. 1979. Maintenance requirements of energy and protein for castrated Japanese native goats. Nutr. Abstr. Rev. 49:1427.

Knowles, F., and J. E. Watkins. 1938. The milk of goats under English conditions. J. Dairy Res. 9:153.

Magee, H. E. 1924. Studies on the metabolism of the ruminant by indirect calorimetry. J. Agric. Sci., Cambridge 14:506.

Magee, H. E. 1932. Studies on the digestion by ruminants. J. Exp. Biol. 9:409.

McDonald, P., R. A. Edwards, and J. F. O. Greenhalgh. 1973. Animal nutrition. Oliver and Boyd, London, England.

Majumdar, B. N. 1960. Studies on goat nutrition. 1. Minimum protein requirement of goats for maintenance—endogenous urinary nitrogen and metabolic faecal nitrogen excretion studies. J. Agric. Sci., Cambridge 54:329.

Morand-Fehr, P., and M. deSimiane. 1977. Proc. Nutr. Goats Mediterran., Granada, Spain.

Morand-Fehr, P., and D. Sauvant. 1980. Composition and yield of goat milk as affected by nutritional manipulation. J. Dairy Sci. 63:1671.

Morgen, A., C. Beger, and G. Fingerling. 1905. Untersuchungen ueber den Einfluss des Nahrungsfettes und einiger anderer Futterbestandteile auf die Milchproduktion. (Studies on the effects of feed fat and some other constituents on milk production.) Landwirtsch. Vers. Stn. 61:1.

Morgen, A., C. Beger, and G. Fingerling. 1906. Weitere Untersuchungen ueber die Wirkung der einzelnen Naehrstoffe auf die Milchproduktion. (Further studies on the effects of single nutrients on milk production.) Landwirtsch. Vers. Stn. 64:93.

Morgen, A., C. Beger, and F. Westhausser. 1907. Untersuchungen ueber den Einfluss des Proteins auf die Milchproduktion, sowie ueber die Beziehungen zwischen Staerkewert und Milchertrag. (About the effect of protein on milk production and relationships between energy contents of feeds and milk yield.) Landwirtsch. Vers. Stn. 66:63.

National Research Council. 1975. Nutrient requirements of sheep. 5th rev. ed. National Academy of Sciences, Washington, D.C.

National Research Council. 1978. Nutrient requirements of dairy cattle. 5th rev. ed. National Academy of Sciences, Washington, D.C.

Olatungi, O. 1974. Digestion and utilization of carbohydrates and energy in West African dwarf sheep. Ph.D. thesis. University of Ibadan, Nigeria.

Opstvedt, J. 1963. Yields from goats on mountain pasture. Sau og Geit. 16:57.

Opstvedt, J. 1967. Studies on the effect of energy level and feed combination on feed intake, milk yield, composition and organoleptic properties. Inst. Anim. Nutr. Agric. Coll. Norway Bull. 134.

Orr, J. B., and H. E. Magee. 1923. The use of indirect calorimetry in ruminants. J. Agric. Sci. 13:447.

Quartermain, A. R. 1973. Aspects of the biological efficiency of the Zambian goat. Proc. 3rd World Conf. Anim. Prod., vol. 2.

Rajpoot, R. L. 1979. Energy and protein in goat nutrition. Ph.D. thesis. Raja Balwant Singh College, Bichpuri (Agra), India.

Rattray, P. V., W. N. Garrett, N. Hirman, and N. N. East. 1974. Energy cost of protein and fat deposition in sheep. J. Anim. Sci. 38:378.

Ritzman, E. G., L. E. Washburn, and F. G. Benedict. 1936. The basal metabolism of the goat. N.H. Agric. Exp. Stn. Tech. Bull. 66.

Sachdeva, K. K., O. P. S. Sengar, S. N. Singh, and I. L. Lindahl. 1973. Studies on goats. I. Effect of plane of nutrition on the reproductive performance of does. J. Agric. Sci., Cambridge 80:375.

Sengar, O. P. S. 1980. Indian research on protein and energy requirements of goats. J. Dairy Sci. 63:1655.

Singh, S. N., and O. P. S. Sengar. 1970. Investigation on milk and meat potentialities of Indian goats, 1965–70. Final Techn. Report Project A7-AH18. Raja Balwant Singh College, Bichpuri (Agra), India.

Tanaka, K., I. Nakajima, and H. Hayashi. 1973. Effect of safflower oil supplement on the plasma lipids of goats. Jpn. J. Zool. 44:19.

Tanaka, K., R. Shimizu, and H. Hayashi. 1973a. The effect of increasing amounts of safflower oil in the diet on the hydrogenation of unsaturated fatty acids in the rumen of goats. Jpn. J. Zool. 44:143.

Tanaka, K., R. Shimizu, and H. Hayashi. 1973b. Effect of dietary supplement of safflower oil or safflower seed on the hydrogenation of unsaturated fatty acids in the rumen of goats. Jpn. J. Zool. 44:212.

Tanaka, K., R. Shimizu, and H. Hayashi. 1973c. Effect of the feeding period of safflower oil on the hydrogenation of unsaturated fatty acids in the rumen of goats. Jpn. J. Zool. 44:278.

Tasaki, I. 1957. Studies on the metabolism of dairy goats. Vol. 6. Comparison of calculation methods for energy metabolism by means of gaseous exchange and nutrient balance. Jpn. J. Zool. 28:37.

Winter, J., and R. Goersch. 1974. Ziegen als Versuchstiere. Ein Beitrag zur Fuetterungsoptimierung. (Goats as experimental animals. A contribution to feeding optimization.) Z. Versuchstierkd. 16:256.

PROTEIN

Akinsoyinu, A. O. 1974. Studies on protein and energy utilization by the West African dwarf goats. Ph.D. thesis. University of Ibadan, Nigeria.

Akinsoyinu, A. O., A. U. Mba, and F. O. Olubaja. 1978. Studies on energy and protein utilization for pregnancy and lactation by the West African dwarf goats in Nigeria. Dairy Sci. Abstr. 40:113.

Astrup, H. N., and J. J. Nedkvitne. 1972. Experimente mit caseinstabilisiertem Fett fuer Wiederkaeuer. (Experiments with casein-stabilized fat for ruminants.) Z. Tierphysiol. Tierernaehr. Futtermittelkd. 30:275.

Becker, O., P. Schulz, and S. Emmerson. 1929. Vergleichende Untersuchungen ueber die Verdauung der Proteine und Kohlehydrate bei Ziegen in Perioden mit und ohne Infusorien. (Comparative studies on the digestion of proteins and carbohydrates in goats with and without protozoa.) Proc. Natl. Acad. Sci. U.S.A. 15:691.

Brody, S. 1945. Bioenergetics and growth. Reinhold, New York.

Broster, W. H. 1972. Protein requirements of cows for lactation, p. 293. *In* W. Lenkeit and K. Breirem, Handbook of animal nutrition, vol. 2. P. Parey, Hamburg, West Germany.

Champredon, C., and R. Pion. 1972a. Effet de la substitution d'urée a du tourteau d'arachide sur la nutrition azotée de la chèvre laitière. (Effect of replacing groundnut meal by urea on nitrogen nutrition of dairy goats.) Ann. Biol. Anim. Biochim. Biophys. 12:229.

Champredon, C., and R. Pion. 1972b. Influence de la teneur en urée du régime sur l'aminoacidémie libre de chèvres en lactation. (Effect of urea content of the diet on free amino acids in blood of lactating goats.) Ann. Biol. Anim. Biochim. Biophys. 12:307.

Champredon, C., and R. Pion. 1972c. Evolution de l'aminoacidémie libre de chèvres au debut de la lactation. (Changes in blood free amino acid levels of dairy goats during the first month of lactation.) Soc. Biol. Fil. 166:378.

Crampton, E. W. 1964. Nutrition to calorie ratios. J. Nutr. 82:253.

Devendra, C. 1967a. Studies on the nutrition of the indigenous goat of Malaysia. I. The maintenance requirements of pen-fed goats. Malays. Agric. J. 46:80.

Devendra, C. 1967b. Studies in the nutrition of the indigenous goat of Malaysia. II. The requirements for live weight gain. Malays. Agric. J. 46:98.

Devendra, C. 1980a. The protein requirements for maintenance of indigenous Kambing Katjang goats in Malaysia. Malays. Agric. Res. Dev. Inst. Res. Bull. (In press).

Devendra, C. 1980b. Milk production in goats compared to buffalo and cattle in humid tropics. J. Dairy Sci. 63:1755.

Devendra, C., and M. Burns. 1970. Goat production in the tropics. Commonw. Bur. Anim. Breed. Genet. Tech. Commun. 19.

Elliott, R. C., and J. H. Topps. 1963. Studies of protein requirements of ruminants. Br. J. Nutr. 17:539.

Fauske, R. 1972. Forsoek med urea og ulik forordning til mjoelkegeit. (Urea and different arrangements of feed for dairy goats.) Nor. Landbrukshoegsk. 401:198.

Ferber, K. E. 1929. Die Veraenderungen der Infusorienzahl im Pansen der Wiederkaeuer im Zusammenhang mit den Veraenderungen des Eiweissumsatzes. (Changes in protozoa numbers in the rumen in relation to changes of protein metabolism.) Z. Tierz. 15:375.

French, M. H., J. Glover, and D. W. Duthie. 1957. The apparent digestibility of crude protein by the ruminant. II. The general equation and some of its implications. J. Agric. Sci., Cambridge 48:379.

French, M. H. 1957. The apparent digestibility of crude protein by the ruminant. III. The application of the general equation to rations containing urea. J. Agric. Sci., Cambridge 48:384.

Fujihara, T., and I. Tasaki. 1973a. Digestibility of casein and starch introduced into rumen and abomasum of goats. Jpn. J. Zool. 44:125.

Fujihara, T., and I. Tasaki. 1973b. Some quantitative aspects of ruminal nitrogen fractions in goats. Jpn. J. Zool. 44:572.

Gallagher, J. R., and M. Shelton. 1972. Efficiencies of conversion of feed to fiber of Angora goats and Rambouillet sheep. J. Anim. Sci. 34:319.

Garrett, W. N., O. H. Mayer, and G. P. Lofgreen. 1959. The comparative energy requirements of sheep and cattle for maintenance and gain. J. Anim. Sci. 18:228.

Gruhn, K., W. Thelemann, and A. Hennig. 1975. Untersuchungen mit markiertem Harnstoff an laktierenden Wiederkaeuern. 4. Der Einbau von 15N-Harnstoff in die basischen Aminosaeuren der Milch von Ziegen. (Studies with labelled urea in lactating ruminants. 4. Incorporation of 15N nitrogen isotope urea into the basic amino acids of goat milk.) Arch. Tierernaehr. 25:629.

Haenlein, G. F. W. 1950. Stoffwechsel und Energiehaushalt der Ziege. (Nutrition and energy metabolism of goats.) Thesis, University of Hohenheim-Stuttgart, West Germany.

Harmeyer, J., and H. Martens. 1980. Aspects of urea metabolism in ruminants with reference to the goat. J. Dairy Sci. 63:1707.

Haryu, T., O. Tanaka, F. Katsumori, R. Tano, J. Akazawa, H. Matsunage, M. Iton, T. Ono, and K. Kameoka. 1975. Soybean meal and urea compared as protein supplements in high-concentrate ration of milking cows and growing goats. Bull. Natl. Inst. Anim. India (Chiba Shi) 29:43.

Haryu, T., R. Tano, M. Itoh, and K. Kameoka. 1975. Factors influencing the utilization of high-grain urea diets by ruminants. II. Effect of protein. Bull. Natl. Inst. Anim. India (Chiba Shi) 29:23.

Hennig, A., K. Gruhn, and W. Thelemann. 1972. Untersuchungen mit markiertem Harnstoff an laktierenden Wiederkaeuern. 2. Die Verteilung von 15N im Ziegenkoerper, im Magen und Darminhalt und die Gesamt-15N-Bilanz zu verschiedenen Zeitpunkten nach der 15N-Applikation. (Studies with labelled urea in lactating ruminants. 2. Distribution of 15N in the body of goats, in gastric and intestinal contents and the total 15N balance at different times after administration of 15N.) Arch. Tierernaehr. 22:509.

Hogan, J. P., and R. H. Weston. 1967. The digestion of two diets of differing protein content but with similar capacities to sustain wool growth. Aust. J. Agric. Res. 18:973.

Honcamp, F., and S. Keudela. 1927. Untersuchungen ueber die Verwertung von Ammoniaksalzen und von Harnstoff als Eiweissersatz fuer die Lebenderhaltung und den Fleischansatz, sowie fuer die Milchbildung. (Studies on utilization of ammonia salts and urea for protein replacement in maintenance, weight gain and milk production.) Z. Tierz. Zuechtungsbiol. 10:1.

Houpt, T. R. 1959. Utilization of blood urea in ruminants. Am. J. Physiol. 197:115.

Hurter, L. R., T. W. Naude, T. F. Adelaar, J. D. Smit, and L. E. Codd. 1972. Ingestion of the plant *Fadogia monticola Robyns* as an additional cause of gousiekte in ruminants. Onderstepoort J. Vet. Res. 39:71.

Huston, J. E., M. Shelton, and W. C. Ellis. 1971. Nutritional requirements of the Angora goat. Tex. Agric. Exp. Stn. Bull. 1105.

Itoh, M., T. Haryu, R. Tano, and K. Iwasaki. 1979. Maintenance requirements of energy and protein for castrated Japanese native goats. Nutr. Abstr. Rev. 49:1427.

Jahn, S. 1956. Verwendung von Amidschnitzeln bei Wiederkaeuern nach Versuchen an laktierenden und graviden Milchziegen. (Utilization of amide-beet pulp in ruminants after studies in milking and pregnant dairy goats.) Ph.D. thesis. University of Halle, East Germany.

Kameoka, K., H. Kawai, M. Itoh, and T. Haryu. 1972. Nitrogen utilization by goats fed semi-purified diets containing diureido isobutane (DUIB). Bull. Natl. Inst. Anim. India 23:428.

Landis, J. 1962. Quantitative aspects of protein metabolism in lactating goats. Int. Z. Vitaminforsch. 32:428.

Lawrow, O., P. Borieta, S. Moltschanowa, and R. Ochotnikowa. 1924. Zur Frage nach der Stickstoffausnutzung des zur Nahrung zugesetzten Harnstoffs bei einem jungen Wiederkaeuer (Boecklein). (Nitrogen utilization from urea supplementation by a young buck.) Biochem. Z. 153:71.

Majumdar, B. N. 1960a. Studies on goat nutrition. 1. Minimum protein requirement of goats for maintenance—endogenous urinary nitrogen and metabolic faecal nitrogen excretion studies. J. Agric. Sci. 54:329.

Majumdar, B. N. 1960b. Studies on goat nutrition. II. Digestible crude protein requirements for maintenance from balance studies. J. Agric. Sci. 54:335.

Mba, A. U., A. O. Akinsoyinu, and F. O. Olubajo. 1974. Studies on comparative utilization of urea and groundnut cake rations by West African dwarf goats. 1. N-balance and growth. Niger. J. Anim. Prod. 1:209.

Mba, A. U., C. P. Egbuiwe, and V. A. Oyenuga. 1975. Nitrogen-balance studies with Red Sokoto (Maradi) goats for the minimum protein requirements. East Afr. Agric. For. J. 40:285.

Morand-Fehr, P., and M. deSimiane. 1977. Proc. Nutr. Goats Mediterran., Granada, Spain.

Morgen, A., C. Beger, and F. Westhausser. 1907. Untersuchungen ueber den Einfluss der nichteiweissartigen Stickstoffverbindungen der Futtermittel auf die Milchproduktion. (About the effects of nonprotein N compounds of feeds upon milk production.) Landwirtsch. Vers. Stn. 6:413.

Morgen, A., C. Beger, and F. Westhausser. 1908. Weitere Untersuchungen ueber den Einfluss der nichteiweissartigen Stickstoffverbindungen der Futtermittel auf die Milchproduktion. (Further studies on effects of nonprotein N compounds of feeds on milk production.) Landwirtsch. Vers. Stn. 68:333.

Morgen, A., C. Beger, and F. Westhausser. 1909. Weitere Untersuchungen ueber die Verwertung der nichteiweissartigen Stickstoffverbindungen der Futtermittel, sowie der Ammonsalze durch das milchgebende Tier. (Utilization of nonprotein N feed compounds and ammonia salts by milking animals.) Landwirtsch. Vers. Stn. 71:1.

Morgen, A., C. Beger, and F. Westhausser. 1910. Untersuchungen ueber die Verwertung der Ammonsalze und der nichteiweissartigen Stickstoffverbindungen der Futtermittel fuer die Lebenderhaltung und Milchbildung, sowie ueber die Frage, ob aus diesen Stoffen unverdauliches Eiweiss gebildet wird. (Utilization of ammonium salts and nonprotein N-compounds for maintenance and milk production and about formation of undigestible protein.) Landwirtsch. Vers. Stn. 73:285.

Morgen, A., A. Schoeler, H. Windheuser, and E. Ohlmer. 1922. Ersatz von Eiweiss durch Harnstoff bei Hammeln und Milchtieren. (Replacement of protein by urea in wethers and milking animals.) Landwirtsch. Vers. Stn. 99:1.

Morgen, A., H. Windheuser, and E. Ohlmer. 1925. Weitere Versuche ueber den Ersatz von Eiweiss durch Harnstoff bei Milchtieren. (Further studies on protein replacement by urea in milking animals.) Landwirtsch. Vers. Stn. 103:1.

National Research Council. 1973 and 1978. Nutrient requirements of horses. 3rd and 4th rev. ed. National Academy of Sciences, Washington, D.C.

National Research Council. 1975. Nutrient requirements of sheep. 5th rev. ed. National Academy of Sciences, Washington, D.C.

National Research Council. 1978. Nutrient requirements of dairy cattle. 5th rev. ed. National Academy of Sciences, Washington, D.C.

Paasch, E. 1925. Fuetterungsversuch an Ziegen mit Ammoniumacetat, Harnstoff und Hornmehl als Eiweissersatz. (Feeding studies with goats to replace protein with ammoniumacetate, urea and hornmeal.) Biochem. Z. 160:352.

Perkins, A. E. 1957. The effect of rations excessively high and extremely low in protein content on dairy cows. Ohio Agric. Exp. Stn. Res. Bull. 799.

Platt, B. S., C. R. C. Heard, and R. J. Stewart. 1964. Experimental protein-calorie deficiency, p. 445. *In* H. N. Munro, Mammalian protein metabolism, vol. 2. Academic Press, New York.

Preston, R. L. 1972. Protein requirements for growing and lactating ruminants. Proc. Univ. Nottingham Nutr. Conf.

Rajpoot, R. L. 1979. Energy and protein in goat nutrition. Ph.D. thesis. Raja Balwant Singh College, Bichpuri (Agra), India.

Reis, P. J., and A. M. Downes. 1971. The rate of response of wool growth to abomasal supplementation of casein. J. Agric. Sci., Cambridge 76:173.

Satter, L. D., and R. E. Roffler. 1975. Nitrogen requirements and utilization in dairy cattle. J. Dairy Sci. 58:1219.

Sengar, O. P. S. 1980. Indian research on protein and energy requirements of goats. J. Dairy Sci. 63:1655.

Singh, S. N. 1976. Nitrogen balances, their significance and limitations. Balwant Vidyapeeth J. Agric. Sci. Res. 4:106.

Singh, N., and V. D. Mudgal. 1978. Studies on endogenous urinary nitrogen, metabolic faecal nitrogen and maintenance requirement for protein in dairy goats. Proc. 20th Int. Dairy Congr., Paris, France.

Singh, S. N., and O. P. S. Sengar. 1970. Investigation on milk and meat potentialities of Indian goats, 1965–70. Final Tech. Report Project A7-AH18. Raja Balwant Singh College, Bichpuri (Agra), India.

Stewart, J. R., M. Shelton, and H. G. Haby. 1971. Nutritional investigations with Angora goats. Tex. Agric. Exp. Stn. Bull. 2933.

Thelemann, W., and K. Gruhn. 1972. Untersuchungen mit markiertem Harnstoff an laktierenden Wiederkaeuern. 1. Die Ausscheidung von 15N aus Harnstoff ueber Milch, Kot und Harn bei Ziegen. (Studies with labelled urea in lactating ruminants. 1. Excretion of 15N from urea in milk, faeces and urine of goats.) Arch. Tierernaehr. 22:205.

Thelemann, W., and A. Hennig. 1973. Untersuchungen mit markiertem Harnstoff an laktierenden Wiederkaeuern. 3. Die Verwertung des Harnstoff-Stickstoffs. (Studies with labelled urea in lactating ruminants. 3. Utilization of urea nitrogen.) Arch. Tierernaehr. 23:175.

Ungerer, E. 1924. Harnstoff und Glykokoll als Eiweissersatz in Versuchen an Milchziegen. (Urea and glycocol as a protein replacement in studies with dairy goats.) Biochem. Z. 147:275.

Varady, J., and J. Harmeyer. 1973. Measurements of nitrogen

recycling in sheep and goats under various conditions. 1. Feeding soybean protein. Inst. Physiol. School Vet. Med., Hannover, West Germany.

Williger, J. 1927. Fuetterungsversuch an Milchziegen mit Glykokoll als Eiweissersatz. (Feeding studies with glycocol to replace protein in dairy goats.) Biochem. Z. 180:156.

Winter, J., and R. Goersch. 1974. Ziegen als Versuchstiere. Ein Beitrag zur Fuetterungsoptimierung. (Goats as experimental animals. A contribution to feeding optimization.) Z. Versuchstierkd. 16:256.

Ziemer, F. H. 1931. Ammoniumbicarbonat neben sauren Zuckerrueben-Diffusionsschnitzeln als Eiweissersatz. (Ammonium bicarbonate besides acidic diffusion beet pulp for protein replacement.) Biochem. Z. 232:352.

MINERALS

Andersson, B. 1955. Observations on the water and electrolyte metabolism in the goat. Acta Physiol. Scand. 33:50.

Barnabas, T., and R. B. Mawal. 1961. Studies on goat's milk during different stages of lactation. II. Distribution of minerals and free amino acids and related compounds during lactation. J. Univ. Poona Sci. Technol. 20:59.

Berg, R. 1941. Die Spurenelemente. (The trace elements.) Z. Ernaehr. 7:33.

Bosworth, A. W., and L. L. Van Slyke. 1916. The soluble and insoluble compounds of goat's milk. J. Biol. Chem. 24:177.

Fingerling, G. 1911. Beitraege zur Frage der Verwertung von Kalk- und Phosphorsaeureverbindungen durch den tierischen Organismus, 1. (Further studies on the utilization of Ca and P compounds by animals, 1.) Landwirtsch. Vers. Stn. 75:1.

Fingerling, G. 1913. Beitraege zur Frage der Verwertung von Kalk- und Phosphorsaeureverbindungen durch den tierischen Organismus, 2. (Further studies on the utilization of Ca and P compounds by animals, 2.) Landwirtsch. Vers. Stn. 80:847.

Gartner, R. J. W., and I. S. Hurwood. 1976. The tannic and oxalic acid content of *Acacia aneura (mulga)* and their possible effects on sulphur and calcium availability. Aust. Vet. J. 52:194.

Gibbons, R. A., B. F. Sansom, and R. Sellwood. 1972. The passage of calcium and strontium across the gut of the anaesthetized goat. J. Physiol. 222:397.

Haardt, E., J. Potocnjak, A. Lopez, and S. Cornejo. 1976. Survey model applied to mineral study in forages for a country (Chile) with a great versatility of ecological systems. Proc. 1st Int. Symp. Feed Comp. Anim. Nutr. Req., Utah Agric. Exp. Stn., Logan, Utah.

Haenlein, G. F. W. 1980. Mineral nutrition of goats. J. Dairy Sci. 63:1729.

Hall, G. A. B. 1976. Phosphorus and trace mineral nutrition of grazing livestock in southern Brazil. Proc. 1st Int. Symp. Feed. Comp. Anim. Nutr. Req., Utah Agric. Exp. Stn., Logan, Utah.

Hamada, T., S. Maeda, and K. Kameoka. 1970. Effects of minerals on formation of color in the rumen epithelium of kids. J. Dairy Sci. 53:588.

Harrold, R. L., and J. D. Nalewaja. 1977. Proximate, mineral and amino acid composition of 15 weed seeds. J. Anim. Sci. 44:389.

Hart, E. B., H. Steenbock, and C. A. Hoppert. 1921. Dietary factors influencing calcium assimilation. I. The comparative influence of green and dried plant tissue, cabbage, orange juice, and cod liver oil on calicum assimilation. J. Biol. Chem. 48:33.

Hart, E. B., H. Steenbock, and C. A. Elvehjem. 1924. Dietary factors influencing calcium assimilation. V. The effect of light upon calcium and phosphorus equilibrium in mature lactating animals. J. Biol. Chem. 62:117.

Hart, E. B., H. Steenbock, S. W. Kletzien, and H. Scott. 1927. Dietary factors influencing calcium assimilation. IX. Further observations on the influence of cod liver oil on calcium assimilation in lactating animals. J. Biol. Chem. 71:271.

Henderson, J. McA., and H. E. Magee. 1926. The effect of ultraviolet light on the calcium and phosphorus metabolism of the lactating animal. Biochem. J. 20:363.

Hennig, A., M. Anke, B. Groppel, and H. Luedke. 1972. Der Einsatz standortspezifischer Mineralstoffgemische in der DDR. (The supply of local specific inorganic mixtures in the German Democratic Republic.) Vol. 1, p. 8. *In* Kongres "Chemia v pol'nohospodarstve," Bratislava, Hungary.

Hunt, C. H., A. R. Winter, J. A. Schultz, and R. D. Miller. 1923. The mineral metabolism of the lactating and dry goat. Am. J. Phys. 66:349.

Hurter, L. R., T. W. Naude, T. F. Adelaar, J. D. Smit, and L. E. Codd. 1972. Ingestion of the plant *Fadogia monticola Robyns* as an additional cause of gousiekte in ruminants. Onderstepoort J. Vet. Res. 39:71.

Lenkeit, E., and M. Schleinitz. 1940. Stoffwechselversuche an Ziegen mit einigen Futtermitteln und deren Einfluss auf den Saeurebasenhaushalt. (Metabolism studies with goats on the effects of certain feeds upon the acid-base balance.) J. Landwirtsch. 87:249.

Lintzel, W. 1931. Der Mineralstoffwechsel. (The mineral metabolism.) *In* E. Mangold, Handbook of nutrition and metabolism of farm animals, vol. III. J. Springer, Berlin, West Germany.

Lintzel, W., and O. Radeff. 1931. Ueber den Eisengehalt und Eisenansatz neugeborener und saeugender Tiere. (Iron content and iron retention of newborn and nursing animals.) Arch. Tierern. Tierz. 6:313.

Macy, I. G., H. J. Kelly, and R. E. Sloan. 1953. The composition of milk. National Research Council Publ. 254, National Academy of Sciences, Washington, D.C.

Mangold, E. 1931. Handbuch der Ernaehrung und des Stoffwechsels der landwirtschaftlichen Nutztiere. (Handbook of the nutrition and metabolism of farm animals.) Vol. I–IV. J. Springer, Berlin, West Germany.

McDowell, L. R., K. R. Fick, R. H. Houser, J. H. Conrad, and J. K. Loosli. 1976. Meeting mineral requirements for grazing livestock in the tropics. Proc. 1st Int. Symp. Feed Comp. Anim. Nutr. Req., Utah Agric. Exp. Stn., Logan, Utah.

Miller, W. J. 1981. Biological value of different sources of inorganic trace elements. Feedstuffs, March 30, p. 20.

National Research Council. 1975. Nutrient requirements of sheep. 5th rev. ed. National Academy of Sciences, Washington, D.C.

National Research Council. 1978. Nutrient requirements of dairy cattle. 5th rev. ed. National Academy of Sciences, Washington, D.C.

National Research Council. 1980. Mineral tolerance of domestic animals. National Academy of Sciences, Washington, D.C.

Parkash, S., and R. Jenness. 1968. Composition and characteristics of goats' milk: A review. Dairy Sci. Abstr. 30:67.

Perdomo, J. T., R. L. Shirley, and C. F. Chicco. 1976. Availability of nutrient minerals in four tropical forages. Proc. 1st Int. Symp. Feed Comp. Anim. Nutr. Req., Utah Agric. Exp. Stn., Logan, Utah.

Reid, R. L., K. Daniel, and J. D. Bubar. 1974. Mineral relationships in sheep and goats maintained on orchardgrass *(Dactylis glomerata)* fertilized with different levels of nitrogen or nitrogen with microelements, over a five-year plan. Proc. 12th Int. Grassland Congr. 2:565.

Reis, P. J., and P. G. Schinkel. 1964. The growth and composition of wool. II. Effects of casein, gelatin and sulfur containing amino acids given *per abomasum*. Aust. J. Biol. Sci. 17:532.

Romagosa Vila, J. A. 1975. Necesidades en oligoelementos en el ganado cabrio. (Micronutrient requirements of goats.) Av. Aliment. Mejora Anim. 16:315.

Sommer, O., and H. Bergmann. 1941. Bericht ueber Fuetterungsversuche mit Heu aus Rauchschaedengebieten des Harzes und des Freistaates Sachsen. (Feeding experiments with hay from smoke damage areas of the Harz mountains and state of Saxony.) Forschungsdienst 12:107.

Stewart, O. 1953. Grazing habits and trace elements. Br. J. Anim. Behaviour 1:116.

Stout, W. L., D. P. Belesky, L. A. Jung, R. S. Adams, and B. L. Moser. 1977. A survey of Pennsylvania forage mineral levels with respect to dairy and beef cow nutrition. Pa. State Coll. Agric. Prog. Rep. 364, Pennsylvania State University, University Park.

Tanabe, S., and K. Kameoka. 1973. Effects of the compositions of major minerals in the purified diet on the performance of preruminant kids. Bull. Natl. Inst. Anim. India 27:17.

Telfer, S. V., and O. Crichton. 1924. Effects of a mineral deficiency in goat milk on growth of young kids. Br. J. Exp. Path. 5:84.

Thomas, B., O. Thompson, S. Oyunaga, and B. Armstrong. 1952. Ash constituents of some herbage plants. Emp. J. Exp. Agric. 20:77.

Thomas, B., and O. Trinder. 1947. Ash components of some moorland plants. Emp. J. Exp. Agric. 15:60.

Ward, G. M. 1966. Potassium metabolism of domestic ruminants—a review. J. Dairy Sci. 49:268.

Wheeler, J. L., D. A. Hedges, and A. R. Till. 1975. A possible effect of cyanogenic glycoside in sorghum on animal requirements for sulfur. J. Agric. Sci., Cambridge 84:377.

Calcium

Anke, M. 1967. Der Mengen- und Spurenelementgehalt des Rinderhaares als Indikator der Kalzium-, Magnesium-, Phosphor-, Kalium-, Natrium-, Eisen-, Zink-, Mangan-, Kupfer-, Molybdaen-, und Kobaltversorgung. (The macro and trace element contents of cattle hair as indicator of the nutritional status of Ca, Mg, P, K, Na, Fe, Zn, Mn, Cu, Mb, and Co.) Arch. Tierernaehr. 17:1.

Fingerling, G. 1911 and 1913. Beitraege zur Frage der Verwertung von Kalk- und Phosphorsaeureverbindungen durch den tierischen Organismus, 1 and 2. (Further studies on the utilization of Ca and P compounds by animals, 1 and 2.) Landwirtsch. Vers. Stn. 75:1; 80:847.

Gibbons, R. A., B. F. Sansom, and R. Sellwood. 1972. The passage of calcium and strontium across the gut of the anaesthetized goat. J. Phys. 222:397.

Hart, E. B., H. Steenbock, and C. A. Hoppert. 1921. Dietary factors influencing calcium assimilation. I. The comparative influence of green and dried plant tissue, cabbage, orange juice, and cod liver oil on calcium assimilation. J. Biol. Chem. 48:33.

Hart, E. B., H. Steenbock, and C. A. Elvehjem. 1924. Dietary factors influencing calcium assimilation. V. The effect of light upon calcium and phosphorus equilibrium in mature lactating animals. J. Biol. Chem. 62:117.

Hart, E. B., H. Steenbock, S. W. Kletzien, and H. Scott. 1927. Dietary factors influencing calcium assimilation. IX. Further observations on the influence of cod liver oil on calcium assimilation in lactating animals. J. Biol. Chem. 71:271.

Henderson, J. McA., and H. E. Magee. 1926. The effect of ultraviolet light on the calcium and phosphorus metabolism of the lactating animal. Biochem. J. 20:363.

Hirch, A., and W. E. Lawrence. 1964. Experimental infection with *Mycobacterium johnei*. 1. The effect of a low calcium diet on the disease in goats. J. Comp. Path. 64:102.

Horst, R. L., and N. A. Jorgensen. 1972. Effect of ammonium chloride on mineral balance and nutrient utilization in lactating and dry goats. J. Dairy Sci. 55:698.

Horst, R. L., and N. A. Jorgensen. 1974. Effect of ammonium chloride on nitrogen and mineral balance in lactating and nonlactating goats. J. Dairy Sci. 57:683.

Macy, I. G., H. J. Kelly, and R. E. Sloan. 1953. The composition of milk. National Research Council Publ. 254, National Academy of Sciences, Washington, D.C.

Parkash, S., and R. Jenness. 1968. Composition and characteristics of goats' milk: A review. Dairy Sci. Abstr. 30:67.

Steenbock, H., and E. B. Hart. 1913. The influence of function on the lime requirements of animals. J. Biol. Chem. 14:59.

Steenbock, H., E. B. Hart, C. A. Elvehjem, and S. W. Kletzien. 1925. Dietary factors influencing calcium assimilation. VI. The antirachitic properties of hays as related to climatic conditions with some observations on the effect of irradiation with ultraviolet light. J. Biol. Chem. 66:425.

Phosphorus

Bonnet, J. A., and A. R. Riera. 1949. Tracing the mineral from the soil to the plant to the animal blood. 3. The effect of phosphorus and lime on the mineral composition of the soil and grass. J. Agric. Univ. Puerto Rico. 33:57.

Bonnet, J. A., A. R. Riera, B. Rivera Brenes, and R. Orlandi. 1949. Effect of urea added to grass fertilized with lime and phosphorus on the nutrition and health of goats. J. Agric. Univ. Puerto Rico. 33:67.

Easter, D. J., S. Patton, and R. D. McCarthy. 1971. Metabolism of phospholipid in mammary gland. I. The supply of phospholipid for milk synthesis in the rat and goat. Lipids 6:844.

Fingerling, G. 1911 and 1913. Beitraege zur Frage der Verwertung von Kalk- und Phosphorsaeureverbindungen durch den tierischen Organismus, 1 and 2. (Further studies on the utilization of Ca and P compounds by animals, 1 and 2.) Landwirtsch. Vers. Stn. 75:1, 80:847.

Jacobson, N. L., K. D. Wiggers, M. H. Wiggers, and G. N. Jacobson. 1976. Phosphorus in dairy cattle and goat nutrition. National Feed Ingredients Assn., W. Des Moines, Iowa.

Luedke, H. 1971. Phosphor und Fruchtarbeit bei Milchkuehen und Ziegen. (Phosphorus and fertility in dairy cattle and goats.) Wiss. Z. Karl-Marx-Univ. Leipzig 20:482.

Macy, I. G., H. J. Kelly, and R. E. Sloan. 1953. The composition of milk. National Research Council Publ. 254, National Academy of Sciences, Washington, D.C.

Otsuka, J., and T. Nakajima. 1958. Effect of starvation and high phosphorus feeding on the bone and liver in goats. Bull. Natl. Inst. Agric. Sci. Jpn. 15:49.

Parkash, S., and R. Jenness. 1968. Composition and characteristics of goats' milk: A review. Dairy Sci. Abstr. 30:67.

Sodium and Chlorine

Andersson, B., L. Eriksson, O. Fernandez, C. G. Kolmodin, and R. Oltner. 1972. Centrally mediated effects of sodium and angiotensin II on arterial blood pressure and fluid balance. Acta Physiol. Scand. 85:398.

Baranowski, S. 1972. Wplyw niedoboru sodu w pokarmie na czynnosc nerek i gruczolu miekowego. (Effect of deficiency of Na in the feed on the function of the kidneys and the mammary gland.) Rozpr. Wyzsza Szk. Roln. Szczecinie 29:32.

Baranowski, S., and R. Siebers. 1973. Wchlanian sodu w nerkach oraz gruczole mlekowym przy braku sodu w pokarmie. (Absorption of sodium in the kidneys and mammary gland on a diet deficient in sodium.) Pol. Arch. Weter. 16:767.

Brigl, P., and H. Windhaeuser. 1929. Ueber Verfuetterung von Natriumchlorat an Schafe und Ziegen. (Feeding NaCl to sheep and goats.) Landwirtsch. Vers. Stn. 109:225.

Peaker, M., and J. L. Linzell. 1974. The effects of oestrus and exogenous oestrogens on milk secretion in the goat. J. Endocrinol. 61:231.

Pino, N., and L. Chiofalo. 1971. Sulle variazioni del tasso sieroematico di Na e di K in *Capra hircus*. Richerche e valutavione biometrico-statistica con riguardo alla eta su un campione di caprini Girgentani in particolari condizioni sperimentali. (On the variations of blood serum levels of Na and K in *Capra hircus* and biometric-statistic evaluation, in relation to the age, on a sample of goats of Girgentana breed in specific experimental conditions.) Zootec. Vita 14:3.

Sachs, J. R., J. C. Ellory, D. L. Kropp, P. B. Dunham, and J. F. Hoffman. 1974. Antibody-induced alterations in the kinetic characteristics of the Na:K (sodium: potassium) pump in goat red blood cells. J. Gen. Physiol. 63:389.

Schellner, G. 1972. Die Wirkung von Natriummangel und Natriumbeifuetterung auf Wachstum, Milch- und Milchfettleistung und Fruchtbarkeit bei Ziegen. (Effect of sodium deficiency and sodium supplements on growth, yields of milk and milk fat and fertility of goats.) Jahrb. Tierernaehr. Fuetterung 8:246.

Magnesium

Razifard, R. 1971. The values of Mg in blood plasma and urine of lactating goat fed on a limited amount of grass. Tehran Univ. J. Vet. Fac. 27:23.

Razifard, R. 1972a. An investigation on the levels of magnesium in blood plasma and milk of the lactating goats. Tehran Univ. J. Vet. Fac. 28:49.

Razifard, R. 1972b. Pattern of blood plasma and urine magnesium in underfed lactating goats. Indian J. Anim. Sci. 42:935.

Potassium

Cline, T. R., M. P. Plumlee, J. E. Christian, and W. V. Kessler. 1969. Effect of potassium perchlorate and sodium chloride given orally upon radioiodine deposition into milk of goats. J. Dairy Sci. 52:1124.

House, W. A., and R. J. Bird. 1975. Magnesium tolerance in goats fed two levels of potassium. J. Anim. Sci. 41:1134.

Sachs, J. R., P. B. Dunham, D. L. Kropp, J. C. Ellory, and J. F. Hoffman. 1974. Interaction of HK (high potassium) and LK (low potassium) goat red blood cells with ouabain. J. Gen. Physiol. 64:536.

Silcock, W. R., and S. Patton. Correlative secretion of protein, lactose and K+ (potassium) in milk of the goat. J. Cell. Physiol. 79:151.

Ward, G. M. 1966. Potassium metabolism of domestic ruminants—a review. J. Dairy Sci. 49:268.

Sulfur

Bhandari, D. S., P. C. Sawhney, and S. P. S. Bedi. 1973. Influence of sulphur and urea on growth and nutrient digestibility in goat kids. Indian J. Anim. Sci. 43:936.

Gartner, R. J. W., and I. S. Hurwood. 1976. The tannic and oxalic acid content of Acacia aneura (mulga) and their possible effects on sulphur and calcium availability. Aust. Vet. J. 52:194.

Reis, P. J., and P. G. Schinkel. 1964. The growth and composition of wool. II. Effects of casein, gelatin and sulfur containing amino acids given *per abomasum*. Aust. J. Biol. Sci. 17:532.

Schober, R., and D. Enkelmann. 1959. Studies on milk formation by in vivo experiments. Incorporation of radioactive sulphate in rumen bacteria, blood plasma and milk proteins. Milchwirtsch. Ztg. 14:484.

Varma, A., and P. C. Sawhney. 1969. Influence of sulphur supplementation on the nutrient utilization in goats on rations containing urea. Indian J. Nutr. Diet. 6:301.

Varma, A., and P. C. Sawhney. 1970. Haematological studies in goats as influenced by supplementation of sulphur on rations containing urea. Indian J. Nutr. Diet. 7:32.

Wheeler, J. L., D. A. Hedges, and A. R. Till. 1975. A possible effect of cyanogenic glycoside in sorghum on animal requirements for sulfur. J. Agric. Sci., Cambridge 84:377.

Iron

Barth, J., and A. L. Mullen. 1972. Adaptation of artificial rumen and simulated abomasal and intestinal fluids in estimating solubility of radionuclides; solubility of 59Fe and secretion into goats' milk. J. Agric. Food Chem. 20:1063.

Hamada, T., S. Maeda, and K. Kameoka. 1970. Effects of minerals on formation of color in rumen epithelium of kids. J. Dairy Sci. 53:588.

Jenness, R. 1980. Composition and characteristics of goat milk: Review, 1968–1979. J. Dairy Sci. 63:1605.

Kolb, E. 1963. The metabolism of iron in farm animals under normal and pathologic conditions. Adv. Vet. Sci. 8:49.

Lintzel, W., and O. Radeff. 1931. Ueber den Eisengehalt und Eisenansatz neugeborener und saeugender Tiere. (Iron content and iron retention of newborn and nursing animals.) Arch. Tierern. Tierz. 6:313.

National Research Council, Committee on Medical and Biologic Effects of Environmental Pollutants. 1979. Iron. University Park Press, Baltimore, Md.

Wallgren, A. 1932. Ueber den Eisengehalt in Milch. (About the iron content of milk.) Acta Paediatr. 12:153.

Iodine

Abdel-Wahab, M. F., and A. M. Osman. 1971. Iodine metabolism in domestic animals in the Sudan usine I_{131}. Endokrinol. 58:198.

Binnerts, W. T., F. W. Lengemann, and C. L. Comar. 1962. Radioiodine metabolism in the goat after simultaneous single and daily administration. J. Dairy Sci. 45:427.

Courth, H. 1931. Untersuchungen ueber den Jodstoffwechsel normaler und thyreoidektomierter Ziegen bei Jodkali und Pflanzenjodfuetterung. (Studies of the iodine metabolism in normal and thyroidectomized goats with KJ and plant iodine feeding.) Biochem. Z. 238:162.

Djurdjevic, D., and F. W. Lengemann. 1970. Effect of perchlorate on the milk-plasma radioiodine ratio of intact and thyroidectomized goats. J. Dairy Sci. 56:808.

Grimmer, O. 1918. Beitraege zur Kenntnis der Milch schilddruesenloser Ziegen. (About the milk of thyroidectomized goats.) Biochem. Z. 88:43.

Honeker, A. 1949. Die Bekaempfung der Kropfbildung bei Ziegen. (The treatment of goiter in goats.) Der Ziegenzuechter 40:24.

Karmarkar, M. G., M. G. Deo, and N. Kochupillai. 1974. Pathophysiology of Himalayan endemic goiter. Am. J. Clin. Nutr. 27:96.

Lengemann, F. W. 1970. Metabolism of radioiodide by lactating goats given iodine-131 for extended periods. J. Dairy Sci. 53:165.

Lengemann, F. W. 1979. Effect of low and high ambient temperatures on metabolism of radioiodine by the lactating goat. J. Dairy Sci. 62:412.

Prakash, P., and D. Sharma. 1972. Thyroxine secretion rate and distribution and uptake of 131-I (iodine) in goats. Indian J. Exp. Biol. 10:20.

Sutphin, M. K., R. W. Hemken, and J. H. Vandersall. 1971. Effect of protein source and iodine level on the metabolism of iodine in goats. J. Dairy Sci. 54:795.

Weiser, S., and J. Zaitschek. 1930. Jodstoffwechselversuch an milchenden Ziegen. (Iodine metabolism studies with milking goats.) Biochem. Z. 217:359.

Weiser, S., and J. Zaitschek. 1932a. Einfluss lang andauernder Jodfuetterung auf den Jodgehalt tierischer Organe. (Effects of long-term iodine feeding upon the iodine content of animal organs.) Biedermanns Zentralbl. Abt. B 3:275.

Weiser, S., and J. Zaitschek. 1932b. Jodtoleranzversuche an Ziegen. (Iodine tolerance experiments with goats.) Biedermanns Zentralbl. Abt. B 3:298.

Weits, J., A. C. Houtman, and J. R. Roborgh. 1953. The effect of an iodinated casein fraction on the milk yield of goats. Netherlands J. Agric. Sci. 1:39.

Copper and Molybdenum

Anke, M., A. Hennig, M. Diettrich, G. Hoffmann, G. Wicke, and D. Pflug. 1971. Resorption, Exkretion und Verteilung von 99Molybdaen nach oraler Gabe an laktierende Wiederkaeuer. (Absorption, excretion and distribution pattern of 99molybdenum after oral administration to lactating ruminants.) Arch. Tierernaehr. 21:505.

Hennig, A., M. Anke, B. Groppel, and H. Luedke. 1974. Secondary copper deficiency in ruminants. Proc. Int. Symp. Trace Element Metabol. Anim. 2d:726.

National Research Council. 1980. Mineral tolerance of domestic animals. National Academy of Sciences, Washington, D.C.

Remy, E. 1932. Zur Frage des Kupfergehaltes der Milch und einiger Milchpraeparate. (About the copper content of milk and some milk products.) Z. Unters. Lebensm. 63:545.

Srivastava, K. B., and P. K. Dwaraknath. 1971. Ceruloplasmin (copper oxidases) activity in the serum of animals. Indian J. Anim. Sci. 41:1044.

Zinc

Groppel, B., and A. Hennig. 1971. Zinkmangel beim Wiederkaeuer. (Zinc deficiency in ruminants.) Arch. Exp. Vet. 25:817.

Miller, W. J., W. J. Pitts, C. M. Clifton, and S. C. Schmittle. 1964. Experimentally produced zinc deficiency in the goat. J. Dairy Sci. 47:556.

National Research Council, Committee on Medical and Biologic Effects of Environmental Pollutants. 1979. Zinc. University Park Press, Baltimore, Md.

Neathery, M. W., W. J. Miller, D. M. Blackmon, F. M. Pate, and R. P. Gentry. 1970. Effect of zinc deficiency in mature goats on feed utilization, hair growth and reproductive characteristics. J. Dairy Sci. 53:682.

Neathery, M. W., W. J. Miller, D. M. Blackmon, F. M. Pate, and R. P. Gentry. 1973. Effects of long term zinc deficiency on feed utilization, reproductive characteristics, and hair growth in the sexually mature male goat. J. Dairy Sci. 56:98.

Papasteriade, A. A. 1973. Study of conditions due to low levels of zinc in sheep and goats in Greece. Ph.D. thesis. Aristotelian University of Thessalonika, Greece.

Schellner, G. 1972. Die Wirkung von Zink- und Manganmangel and Zink- und Manganzufuetterung auf Wachstum, Milch- und Milch-fettleistung und Fruchtbarkeit beim Wiederkaeuer. (Effect of deficiencies of zinc and manganese and supplements of zinc and manganese on growth, yields of milk and milk fat, and fertility in ruminants.) Jahrb. Tierernaehr. Fuetterung. 8:137.

Schulze, A., and K. M. Ustdal. 1975. Moegliche Ursachen der Alopezie tuerkischer Angora-Ziegen. (Possible causes of alopecia in Turkish Angora goats.) Berl. Muench. Tieraerztl. Wochenschr. 88:66.

Sonwane, S. N., and S. P. Arora. 1975. Influence of dietary levels of zinc on its secretion in milk. Indian J. Dairy Sci. 28:99.

Manganese

Anke, M., A. Hennig, G. Hoffmann, B. Groppel, H. Luedke, and M. Gruen. 1972. Resorption, Exkretion und Inkorporation von 54Mangan aus Manganammoniumphosphat durch Gefluegel und Wiederkaeuer. (Absorption, excretion and incorporation of 54Manganese from manganese ammonium phosphate by poultry and ruminants.) Arch. Tierernaehr. 22:347.

Anke, M., B. Groppel, W. Reissig, H. Luedke, M. Gruen, and G. Dittrich. 1973. Manganmangel beim Wiederkaeuer. 3. Manganmangelbedingte Fortpflanzungs-, Skelett- und Nervenstoerungen bei weiblichen Wiederkaeuern und ihren Nachkommen. (Manganese deficiency in ruminants. 3. Disorders of reproduction, skeleton and nerves caused by manganese deficiency in female ruminants and their progeny.) Arch. Tierernaehr. 23:197.

Anke, M., A. Hennig, B. Groppel, G. Dittrich, M. Gruen, and G. Schellner. 1973. Manganmangel beim Wiederkaeuer. 4. Der

Einfluss des Manganmangels auf den Gehalt neugeborener Laemmer an Fett, Protein, Mangan, Asche, Kalzium, Phosphor, Zink und Kupfer. (Manganese deficiency in ruminants. 4. Effect of manganese deficiency on fat, protein, manganese, ash, calcium, phosphorus, zinc and copper contents of newborn kids.) Arch. Tierernaehr. 23:213.

Anke, M., B. Groppel, and M. Gruen. 1973. Manganmangel beim Wiederkaeuer. 5. Der Einfluss des Manganmangels auf den Mengen- und Spurenelementgehalt erwachsener weiblicher und maennlicher Ziegen. (Manganese deficiency in ruminants. 5. Effect of manganese deficiency on major and trace element contents of adult male and female goats.) Arch. Tierernaehr. 23:483.

Groppel, B. 1969. Vergleichende Untersuchungen ueber den Mangan-, Zink- und Cadmiumstoffwechsel kleiner Wiederkaeuer bei unterschiedlicher Versorgung mit diesen Elementen. (Comparative studies of the Mn, Zn and Cd metabolism of small ruminants during different supplies with these elements.) Ph.D. thesis. University of Jena, East Germany.

Hennig, A., M. Anke, B. Groppel, H. Luedke, W. Reissig, G. Dittrich, and M. Gruen. 1972. Manganmangel beim Wiederkaeuer. 1. Der Einfluss des Manganmangels auf die Lebendmasseentwicklung. (Manganese deficiency in ruminants. 1. Effect of manganese deficiency on liveweight.) Arch. Tierernaehr. 22:601.

Schellner, G. 1972. Die Wirkung von Zink- und Manganmangel und Zink- und Manganzufuetterung auf Wachstum, Milch- und Milchfettleistung und Fruchtbarkeit beim Wiederkaeuer. (Effect of deficiencies of zinc and manganese and supplements of zinc and manganese on growth, yields of milk and milk fat, and fertility in ruminants.) Jahrb. Tierernaehr. Fuetterung. 8:137.

Other Minerals

Anke, M., A. Hennig, M. Gruen, M. Partschefeld, B. Groppel, and H. Luedke. 1977. Nickel, an essential trace element. I. The supply of nickel as affecting the live weight gains, food consumption and body composition of growing pigs and goats. Arch. Tierernaehr. 27:25.

Askew, H. 1947. The effectiveness of small applications of cobalt sulphate. N.Z. J. Sci. Techol. A 5:186.

Barth, J., A. N. Mikalis, J. Y. Harris, and B. H. Bruckner. 1969. Evaluation of clays as binding agents for reduction of radionuclides in milk. Effect of Belle Fourche bentonite on excretion of 134Cs in lactating goats. J. Agric. Food Chem. 17:1347.

Barth, J., and A. L. Mullen. 1972. Adaption of artificial rumen and simulated abomasal and intestinal fluids in estimating solubility of radionuclides; solubility of 59Fe and secretion into goats' milk. J. Agric. Food Chem. 20:1063.

Hazzard, D. G. 1969. Percent cesium-134 and strontium-85 in milk, urine, and feces of goats on normal and verxite-containing diets. J. Dairy Sci. 52:990.

Hebert, D. M., and I. M. Cowan. 1971. White muscle disease in the mountain goat. J. Wildl. Manage. 35:752.

Hennig, A., M. Anke, M. Gruen, M. Partschefeld, and B. Groppel. 1978. Neuere Spurenelemente in der Ernaehrung der landwirtschaftlichen Nutztiere und Laboratoriumstiere. (Newer trace elements in the nutrition of farm and laboratory animals.) Proc. 3rd World Congr. Anim. Prod., Madrid, Spain.

Hosseinion, M., T. T. Bazargani, J. Nahani, H. Mohammadiha, and A. Owlia. 1972. Selenium poisoning in a mixed flock of sheep and goats in Iran. Trop. Anim. Health Prod. 4:173.

Howe, M., J. McGee, and F. W. Lengemann. 1972. Transfer of inorganic mercury to milk of goats. Nature (London) 237:516.

Miessner, H. 1927. Fuetterungsversuche mit Blei. (Feeding trials with lead.) Tieraerztl. Wochenschr. 35:297.

National Research Council. 1975. Nutrient requirements of sheep. 5th rev. ed. National Academy of Sciences, Washington, D.C.

Panchenko, I. Ya., I. S. Irvin, N. I. Burov, and A. N. Sirotkin. 1974. Nakoplenie strontsiya-90 v organizme razlichnykh zhivotnykh. (Accumulation of strontium 90 in different animals.) Skh. Biolog. 9:124.

Philipson, A. T., and H. Mitchell. 1952. The administration of cobalt to lambs on a low cobalt diet. Br. J. Nutr. 6:291.

Platten, O. 1951. Cobalt requirement of the goat, p. 12. *In* Br. Goat Soc. Yearb.

Schreckhise, R. G., and R. L. Watters. 1969. Internal distribution and milk secretion of polonium-210 after oral administration to a lactating goat. J. Dairy Sci. 52:1867.

Sell, J. L., and K. L. Davison. 1973. Mercury-203 in milk and tissues of cow and goat following intraruminal injection of methyl mercuric (labeled 203) chloride. J. Dairy Sci. 56:671.

Sutton, W. W. 1976. Biological transfer of plutonium via in vivo labeled goats' milk. Proc. Symp. Environ. Protection Agency, Las Vegas, Nev.

VITAMINS

Asdell, H. 1941. Transference of dietary vitamin E into goats' milk. *In* Br. Goat Soc. Yearb.

Bhargrat, O., and P. Devi. 1944. Vitamin-B inhibitors in cereals and pulses. Indian J. Med. Res. 32:1.

Caldas, A. D. 1961. Spontaneous vitamin A deficiency in goats. Biologico 27:266.

Caola, G., A. Ferlazzo, and G. Micali. 1972. La citremia nella capra gravida. (Citric acid in blood of pregnant goats.) Atti. Soc. Ital. Sci. Vet. 26:211.

Chanda, R., H. M. Clapham, M. L. McNaught, and E. C. Owen. 1951. The digestibility of carotene by the cow and the goat as affected by thyroxine and thiouracil. Biochem. J. 50:95.

Chanda, R., and E. C. Owen. 1952. The vitamin-A content of the milk of cows and goats on a comparable diet. Biochem. J. 51:404.

Coleman, R., and H. Riemann. 1976. A reminder: Deficiency of folic acid in goat milk. J. Pediatr. 88:900.

Davis, O. 1942. Coccidiosis as a cause of vitamin-A deficiency. Nature 170:849.

DeLuca, H. F. 1974. Vitamin-D: The vitamin and the hormone. Fed. Proc. 33:2211.

DeRuyter de Wildt, J. C., and E. Brouwer. 1929. Over de behoefte van jonge geiten aan vitamine C. (About the need of young goats for vitamin C.) Landbouwk. Onderzoeking. Rijkslandbouwproefstat. 35:88.

Eveleth, D. F., D. W. Bolin, and A. I. Goldsby. 1949. Experimental avitaminosis A in sheep. Am. J. Vet. Res. 10:250.

Ford, J. E., G. S. Knaggs, D. N. Salter, and K. J. Scott. 1972. Folate nutrition in the kid. Br. J. Nutr. 27:571.

Frier, H. I., E. J. Gorgacz, R. C. Hall, Jr., A. M. Gallina, J. E. Rousseau, Jr., H. D. Eaton, and S. W. Nielsen. 1974. Formation and absorption of cerebrospinal fluid in adult goats with hypo- and hypervitaminosis A. Am. J. Vet. Res. 35:45.

Goto, M., K. Okamoto, S. Kawamura, and M. Takahashi. 1975. Formation of vitamin B group and the amino acid pattern of microbial protein in the rumen of the goat fed with protein-free diet containing ureidocompound and hay. Bull. Fac. Agric. Meiji Univ. 33:7.

Holmes, A. D., H. G. Lindquist, and E. K. Greenwood. 1945. Variations in fat, ascorbic acid and riboflavin content of goats' milk. J. Dairy Sci. 28:853.

Honeker, A. 1949. Vitamine und Vitaminmangelkrankheiten bei Ziegen. (Vitamins and vitamin deficiency diseases.) Der Ziegenzuechter 40:61.

Jenness, R. 1980. Composition and characteristics of goat milk: Review, 1968–1979. J. Dairy Sci. 63:1605.

Kuroki, M., M. Yamaguchi, and I. Iwamura. 1956. Observation and discussion on the effect of vitamin B_{12}-administration in goats on the secretion of milk. Bull. Fac. Agric. Univ. Miyazaki 1:171.

Linzell, J. L., T. B. Mepham, and M. Peaker. 1976. The secretion of citrate into milk. J. Physiol. 260:739.

Majumdar, B. N., and B. N. Gupta. 1960. Studies on carotene metabolism in goats. Indian J. Med. Res. 48:388.

Muth, O. H., J. E. Oldfield, L. F. Remmert, and J. R. Schubert. 1958. White muscle disease (myopathy) in lambs and calves. Effects of selenium and vitamin E on lambs. Science 128:1090.

National Research Council. 1975. Nutrient requirements of sheep. 5th rev. ed. National Academy of Sciences, Washington, D.C.

National Research Council. 1978. Nutrient requirements of dairy cattle. 5th rev. ed. National Academy of Sciences, Washington, D.C.

Pavel, J., and J. Zak. 1969. Die Veraenderungen des Gehaltes einiger hydrophiler Vitamine in Beziehung zur Entfaltung der Mikroflora im Pansensaft bei Wiederkaeuern bei normaler und halbsynthetischer Diaet. (Changes in the content of some water-soluble vitamins in relation to the development of the microflora in the ruminal juice of ruminants on normal and semisynthetic diets.) Arch. Exp. Veterinaermed. 23:65.

Peaker, M., and J. L. Linzell. 1975. Citrate in milk: A harbinger of lactogenesis. Nature 253:464.

Schmidt, H. 1941. Vitamin A deficiencies in ruminants. Am. J. Vet. Res. 2:373.

Wites, H. 1942. Vitamin-D inhibitor in hay. Nature 170:871.

Zannelli, C. 1958. Influenza dell'alimentazione mista ed esclusiva con sottoprodotti dell'industria agrumaria sulla secrezuone lattea della capra. 5. Influenza sulla vitamina C. (Effect of mixed feeding and feeding only on by-products of the citrus industry on milk secretion in the goat. 5. Effects of vitamin C.) Boll. Soc. Ital. Biol. Sper. 34:41.

WATER

Anderson, B., and B. Larsson. 1961. Influence of local temperature changes in the preoptic area and rostral hypothalamus on the regulation of food and water intake. Acta Physiol. Scand. 52:75.

Asuncion, P. P. 1939. A comparative study of the forage and water consumption of sheep and goats. Philipp. Agric. 28:576.

Bergman, H. 1932. Der Wasserbedarf der Ziegen. (Water requirements of goats.) Der Ziegenzuechter 19:146.

Brown, G. D., and J. J. Lynch. 1972. Some aspects of the water balance of sheep at pasture when deprived of drinking water. Aust. J. Agric. Res. 23:669.

Choshniak, I., and A. Shkolnik. 1978. The rumen as a protective osmotic mechanism during rapid rehydration in the black Bedouin goat. Proc. Alfred Benzon Symp., Munkegaard, Copenhagen, Denmark.

Dawson, T. J., M. J. S. Denny, E. M. Russell, and B. Ellis. 1975. Water usage and diet preferences of free ranging kangaroos, sheep and feral goats in the Australian arid zone during summer. J. Zool. 177:1.

Devendra, C. 1967. Studies in the nutrition of the indigenous goat of Malaya. 4. The free-water intake of pen-fed goats. 5. Food conversion efficiency, economic efficiency and feeding standards for goats. Malays. Agric. J. 46:191.

Dunson, W. A. 1974. Some aspects of salt and water balance of feral goats from arid islands. Am. J. Physiol. 226:662.

Goatcher, W. D., and D. C. Church. 1970. Taste responses of ruminants. J. Anim. Sci. 31:373.

Khan, M. S., P. K. Ghosh, and T. O. Sasidharan. 1978. Effect of acute water restriction on plasma proteins and on blood and urinary electrolytes in Barmer goats of the Rajasthan desert. J. Agric. Sci., Cambridge 91:42.

Leitch, O., and J. S. Thomson. 1944. The water economy of farm animals. Nutr. Abstr. Rev. 14:197.

Lynch, J. J., G. D. Brown, P. F. May, and J. B. Donnelly. 1972. The effect of withholding drinking water on wool growth and lamb production of grazing Merino sheep in a temperate climate. Aust. J. Agric. Res. 23:659.

Macfarlane, W. V., and B. Howard. 1972. Comparative water and energy economy of wild and domestic animals. Symp. Zool. Soc. London 31:261.

Maloiy, G. M. O., and C. R. Taylor. 1971. Water requirements of African goats and haired-sheep. J. Agric. Sci., Cambridge 77:203.

Maloiy, G. M. O. 1973. Water metabolism of East African ruminants in arid and semi-arid regions. Z. Tierz. Zuechtungsbiol. 90:219.

Morand-Fehr, P., and D. Sauvant. 1978. Caprins, chap. 15. *In* Alimentation des ruminant. (Ruminant nutrition.) Inst. Natl. Rech. Agron., Paris, France.

Schmidt-Nielson, K. 1964. Desert animals' physiological problem of heat and water. Clarendon Press, Oxford, England.

Schoen, A. 1968. Studies on the water-balance of the East African goat. E. Afr. For. J. 34:256.

Shkolnik, A., A. Borut, and J. Chosniak. 1972. Water economy of the Bedouin goat. Symp. Zool. Soc. London 31:229.

Shkolnik, A., A. Borut, I. Chosniak, and A. Maltz. 1974. Water economy and drinking regime of the Bedouin goat. Proc. Symp. Ecol. Res. Developm. Arid Zones Winter Precip., Volcanic Center, Bet-Dagan, Israel, 39.

Shkolnik, A., A. Borut, and I. Choshniak. 1979. Water economy of Bedouin goats. Symp. Zool. Soc. London 31:229.

Shkolnik, A., E. Maltz, and S. Gordin. 1980. Desert conditions and goat milk production. J. Dairy Sci. 63:1749.

Vyas, U. K., P. L. Arya, and S. N. Sharma. 1978. Studies on experimentally produced uremia in sheep and goats. Indian J. Anim. Sci. 91:42.

Wildenhahn, V. 1967. Ueber die Gewoehnung von Ziegen an eiskaltes Traenkwasser. (Adaptation of goats to ice-cold water.) Arch. Tierernaehr. 17:135.

TABLE OF NUTRIENT REQUIREMENTS

Garrett, W. N., O. H. Mayer, and G. P. Lofgreen. 1959. The comparative energy requirements of sheep and cattle for maintenance and gain. J. Anim. Sci. 18:228.

Haenlein, G. F. W. 1978. Dairy goats do well on free choice feeding. Hoard's Dairyman 123:1194.

HERBAGE AND BROWSE UTILIZATION

Ademosun, A. A. 1970a. Nutritive evaluation of Nigerian forages. 1. Digestibility of *Pennisetum purpureum* by sheep and goats. Niger. Agric. J. 7:19.

Ademosun, A. A. 1970b. Nutritive evaluation of Nigerian forages. 2. The effect of stage of maturity on the nutritive value of *Stylosanthes gracilis*. Niger. Agric. J. 7:164.

Adeyanju, S. A., D. B. A. Ogutuga, J. O. Ilori, and A. A. Adegbola. 1975. Cocoa husk in maintenance rations for sheep and goats in the tropics. Nutr. Rep. Int. 11:351.

Askins, G. D., and E. E. Turner. 1972. A behavioral study of Angora goats on West Texas range. J. Range Manage. 25:82.

Asuncion, P. P. 1939. A comparative study on the forage and water consumption of sheep and goats. Philipp. Agric. 28:576.

Axtmayer, J. H., C. J. Asenjo, and D. H. Cook. 1938. The nutritive value of some forage crops of Puerto Rico. I. Grasses. J. Agric. Univ. Puerto Rico. 22:95.

Banerjee, G. C., L. Mandal, A. Mukherjee, and D. N. Roy. 1973. Nutritive value of *Pennisetum pedicullatum* grass for black Bengal goat. Indian J. Anim. Health 12:143.

Baumgardt, B. R., W. J. Byer, H. F. Jumah, and C. R. Drueger. 1964. Digestion in the steer, goat, and artificial rumen as measures of forage nutritive value. J. Dairy Sci. 47:160.

Bawa, M. S. 1950. Utility of date fruit as a feed for goat. Indian Farming. 11:328.

Becker, M., M. V. Schleinitz, and E. Lagneau. 1939. Versuche ueber kuenstlich getrocknete Gruenfuttermittel. (Studies of artificially dried forages.) Zuechtungskunde 14:51.

Bell, F. R. 1959. Preference thresholds for taste discrimination in goats. J. Agric. Sci. 52:125.

Bergmann, H. 1947. Kastanien als Ziengenfutter. (Horse chestnuts as goat feed.) Mitteilungsbl. Ziegenzuechter 1:7.

Bhandari, D. S., and M. L. Gupta. 1973. Utilization of ardu leaves (*Ailanthus excelsa*) as a supplement for poor quality roughage. Indian Vet. J. 50:809.

Bissell, H. D., and W. C. Weir. 1957. The digestibilities of interior live oak and chamise by deer and sheep. J. Anim. Sci. 16:476.

Bredon, R. M., and B. Marshall. 1962. The chemical composition and nutritive value of some common grasses in Uganda. II. The comparison of chemical composition and nutritive values of grasses throughout the year, with special reference to the later stages of growth. Trop. Agric. (Trinidad) 39:13.

Butterworth, M. H. 1967. The digestibility of tropical grasses. Nutr. Abstr. Rev. 37:349.

Butterworth, M. H., C. G. Groom, and P. N. Wilson. 1961. The intake of Pangola grass (*Digitaria decumbens, Stent.*), under wet- and dry-season conditions in Trinidad. J. Agric. Sci., Cambridge 56:407.

Byer, W. J., H. F. Jumah, B. R. Baumgardt, and R. P. Niedermeier. 1961. Digestion in the steer, goat and artificial rumen as measures of the nutritive value of alfalfa harvested at four stages of maturity. Proc. No. P-3, Annu. Meet. Am. Dairy Sci. Assoc., Univ. Wisc., Madison.

Campbell, Q. P., J. P. Ebersoehn, and H. H. von Broembsen. 1962. Browsing by goats and its effects on the vegetation. Herb. Abstr. 32:273.

Carrera, C., and J. C. Blake. 1968. Plantas aprovechadas por el ganado caprino en una zona de matorral desértico y su analisis quimico proximal. (Plants eaten by goats in a bushy desert zone and their proximate composition.) Rev. Mex. Prod. Anim. 1:17.

Castle, E. J. 1956a. The rate of passage of foodstuffs through the alimentary tract of the goat. I. Studies on adult animals fed hay and concentrates. Br. J. Nutr. 10:15.

Castle, E. J. 1956b. The rate of passage of foodstuffs through the alimentary tract of the goat. II. Studies on growing kids. Br. J. Nutr. 10:115.

Chenost, M. 1972a. Observations préliminaires sur la comparaison du potentiel digestif et de l'appétit des caprins et des ovins en zone tropicale et en zone tempérée. (Comparison of digestive potential and appetite of goats and sheep in tropical and temperate zones.) Ann. Zootech. 21:107.

Chenost, M. 1972b. Observations préliminaires sur les variations saisonnières de la quantité d'aliment ingérée par les caprins en milieu tropical humide. (Seasonal variations in amount of feed eaten by goats in a humid tropical environment.) Ann. Zootech. 21:113.

Chenost, M. 1972c. Influence de l'apport d'aliment concentré sur la digestibilité et les quantités ingérées de rations à base de Pangola (*Digitaria decumbens*) distribué en vert à des chèvres laitières et à des boucs. (Influence of concentrate allowance on digestibility and intake of diets based on Pangola grass (*Digitaria decumbens*) given green to lactating and male goats.) Ann. Zootech. 21:121.

Chenost, M. 1973. La valeur alimentaire de quatre graminées et d'une légumineuse tropicales et ses facteurs de variation. (Nutritive value of four grasses and a tropical legume and their variation factors.) Fourrages 54:87.

Chenost, M., and P. Bousquet. 1974. Exploitation en vert du Pangola pour la production de lait par des chèvres. 1. Alimentation complémentaire suivant la qualité du forrage. (Use of green Pangola grass for milk production in goats. 1. Supplementary feed according to the quality of the forage.) Ann. Zootech. 23:1.

Chenost, M., P. Despois, A. Grude, G. Saminadin, and A. Paul-Urbain-Georges. 1975. La valeur alimentaire du Pangola (*Digitaria decumbens, Stent.*) et ses facteurs de variation en zone tropicale humide. (Factors affecting feeding value of Pangola grass (*Digitaria decumbens, Stent.*) in a humid tropical climate.) Ann. Zootech. 24:327.

Cordova, F. J., and J. D. Wallace. 1975. Nutritive value of some browse and forb species. Proc. West. Sect. Am. Soc. Anim. Sci. 26:160.

Cory, V. L. 1927. Activity of livestock on the range. Tex. Agric. Exp. Stn. Bull. 367.

Coblenz, B. E. 1977. Some range relationships of feral goats on Santa Catalina Island, California. J. Range Manage. 30:415–419.

Devendra, C. 1967a. Studies on the nutrition of the indigenous goat of Malaya. 3. The requirement for liveweight gain. Malays. Agric. J. 46:98.

Devendra, C. 1967b. Studies in the nutrition of the indigenous goat of Malaya. 4. The free-water intake of pen fed goats. Malays. Agric. J. 46:191.

Devendra, C. 1971. The comparative efficiency of feed utilization of ruminants in the tropics. Trop. Sci. 13:123.

Devendra, C. 1973. Species differences in the digestibility of Napier grass (*Pennisetum purpureum*) by goats and sheep. Proc. 3rd World Conf. Anim. Prod., Melbourne, Australia.

Devendra, C. 1975. The intake and digestibility of Napier grass (*Pennisetum purpureum*) at four, five and six weeks of growth by goats and sheep in Trinidad. Turrialba 25:226.

Devendra, C. 1977a. Studies in the intake and digestibility of two varieties (*Serdang* and *Coloniao*) of Guinea grass (*Panicum maximum*) by goats and sheep. 1. Long grass. Malays. Agric. Res. Developm. Inst. Res. Bull. 5:91.

Devendra, C. 1977b. Studies in the intake and digestibility of two varieties (*Serdang* and *Coloniao*) of Guinea grass (*Panicum maximum*) by goats and sheep. 2. Chopped and pelleted grass fed to sheep. Malays. Agric. Res. Developm. Inst. Res. Bull. 5:110.

Devendra, C. 1978. The digestive efficiency of goats. World Rev. Anim. Prod. 14:9.

Devendra, C. 1980. The potential of sheep and goats in developing countries. J. Anim. Sci. 51:461.

Djafar, M. I., and F. Y. Chin. 1972. Nutritive values of four grasses grown in Malaysia. Malays. Agric. Res. 1:110.

Edwards, L. C. 1948. Some notes on the food of goats in a semi-arid area. E. Afr. Agric. J. 13:221.

Ehrlein, H. J., and H. Hill. 1970. Einfluesse des Labmagen- und Duodenalinhaltes auf die Motorik des Wiederkauermagens. (The influence of the contents of the abomasum and the duodenum on the motility of the ruminant stomachs.) Zentralbl. Veterinaermed. Reihe A 17:498.

ElHag, G. A. 1976. A comparative study between desert goat and sheep efficiency of feed utilization. World Rev. Anim. Prod. 12:43.

Fehr, P. M. 1971. Utilisation des fourrages granulés par la chèvre laitiére. (Utilization of pelleted forage by dairy goats.) Ann. Zootech. 20:305.

Fehr, P. M., and R. Disset. 1971. La luzerne agglomerée pour les chèvres. (Compressed alfalfa for goats.) Rev. Elevage 49:77.

Fingerling, G. 1905. Untersuchungen ueber den Einfluss von Reizstoffen auf die Futteraufnahme, Verdaulichkeit, Milchsekretion bei reizlosem und normalem Futter. (Effects of taste substances on feed intake, digestibility and milk secretion with bland and normal feed.) Landwirtsch. Vers. Stn. 62:11.

Fingerling, G. 1906. Neuer Apparat zur getrennten Auffangung von Kot und Harn bei kleineren weiblichen Tieren (Ziegen und Schafen). (Apparatus for separate collection of feces and urine in small female animals—goats and sheep. Z. Biol. 47:72.

Fingerling, G. 1907. Weitere Untersuchungen ueber den Einfluss von Reizstoffen auf die Milchsekretion. (More studies on taste substances upon milk secretion.) Landwirtsch. Vers. Stn. 67:253.

Fingerling, G. 1909. Weitere Untersuchungen ueber den Einfluss von Reizstoffen auf die Milchsekretion. (Further studies on the effects of taste substances upon milk secretion.) Landwirtsch. Vers. Stn. 71:373.

Fonolla, J., R. Sanz, and A. Gonzalez Moles. 1972. Influencia de la granulación sobre la digestibilidad y rendimiento nutritivo de la harina de alfalfa deshidratada en cabras de raza Granadina. (Effect of pelleting on digestibility and feeding value of dried alfalfa meal for Granada goats.) Av. Aliment. Mejora Anim. 13:343.

Fraps, G. S. 1924. Digestion experiments with oat by-products and other feeds. Tex. Agric. Exp. Stn. Bull. 315.

Fraps, G. S., and V. L. Cory. 1940. Composition and utilization of range vegetation of Sutton and Edwards counties. Tex. Agric. Exp. Stn. Bull. 586.

French, M. H. 1957. Nutritional value of tropical grasses and fodders. Herb. Abstr. 27:1.

Gartner, R. J. W., and I. S. Hurwood. 1976. The tannic and oxalic acid content of *Acacia aneura (Mulga)* and their possible effects on sulphur and calcium availability. Aust. Vet. J. 52:194.

Geoffroy, F. 1974. Etudé comparée du comportement alimentaire et merycique de deux petits ruminants: La chèvre et le mouton. (Comparison of feeding and ruminating behavior in two small ruminants: Goats and sheep.) Ann. Zootech. 23:63.

Gihad, F. A. 1976. Intake, digestibility and nitrogen utilization of tropical natural grass hay by goats and sheep. J. Anim. Sci. 43:879.

Gihad, F. A., T. M. el-Bedawy, and A. Z. Mehrez. 1980. Fiber digestibility by goats and sheep. J. Dairy Sci. 63:1701.

Gomez-Guillamon, L., G. Varela, and J. Boza. 1961. Valor leche del higo chumbo desecado en cabras de raza granadina. (Milk production value of dried prickly pear in goats of Granada breed.) Zootechn. 10:190.

Gordon, O., and W. Tribe. 1937. Importance of frequent feeding. Br. J. Nutr. 6:89.

Griego, R. R., and J. C. Malechek. 1975. Sheep and goat grazing responses in the pre-Sahara. Proc. West. Soc. Anim. Sci. 26:151.

Gupta, J. H., S. K. Pramanik, and B. N. Majumdar. 1967. A comparison of the nutritive value of hybrid Napier, Napier and Jowar grasses with or without starch supplementation. J. Agric. Sci., Cambridge 68:135.

Gusejnov, D. Ja. 1962. O laktogennom dejstvii bojarysnika. (Lactogenic effect of hawthorns.) Veterinarija (Moscow) 39:61.

Hamada, T. 1973. Feed utilization and anatomical adaptation of kids after removal of compartments of forestomach or omasal laminae. J. Dairy Sci. 56:473.

Hamada, T., H. Hayakawa, and K. Kameoka. 1973. Feed values, ruminal decomposition, and ad libitum intakes of alfalfa hay cube and Italian ryegrass hay by the goat. Bull. Natl. Inst. Anim. India 27:11.

Hamada, T., T. Okubo, H. Hayakawa, and K. Kameoka. 1974. Estimation of energy and crude protein digestibilities of growing alfalfa, perennial ryegrass, and tall fescue by the continuous digestion experiment. Nippon Sochi Gakkai-Shi 20:22.

Heffron, C. L., J. T. Reid, W. M. Haschek, A. K. Furr, T. F. Parkinson, C. A. Bache, W. H. Gutenmann, L. E. St. John, Jr., and D. J. Lisk. 1977. Chemical composition and acceptability of aquatic plants in diets of sheep and pregnant goats. J. Anim. Sci. 45:1166.

Hjeljord, O. 1973. Mountain goat forage and habitat preference in Alaska. J. Wildl. Manage. 37:353.

Hofmeyr, H. S., D. M. Joubert, F. J. G. Badenhorst, and G. J. van D. Steyn. 1965. Adaptability of sheep and goats to a South African tropical environment. Proc. S. Afr. Soc. Anim. Prod. 4:191.

Hoppe, P. P., S. A. Ovortrup, and M. H. Woodford. 1977. Rumen fermentation and food selection in East African sheep, goats, Thompson's gazelle, Grant's gazelle and impala. J. Agric. Sci. 89:129.

Hornby, H. E., and H. J. van Rensburg. 1948. The place of goats in Tanganyika farming systems. I. *In* deciduous bushland formation. E. Afr. Agric. J. 14:94.

Hossain, W. 1960. Evaluation of pipal leaves as a feed for goats. Agric. Pak. 11:139.

Huss, D. 1971. Goat responses to use of shrubs in *Wildlife Shrubs . . . Their Biology and Utilization*. An international symposium, Utah State Univ., Logan, 7 pp.

Huston, J. E. 1976. Relative digestive capacities of cattle, sheep, goats and deer for a common feed. Tex. Agric. Exp. Stn. Bull. 3400:36.

Huston, J. E. 1978. Forage utilization and nutrient requirements of the goat. J. Dairy Sci. 61:988.

Jang, S., and B. N. Majumdar. 1962. A study of comparative digestibilities in different species of ruminants. Ann. Biochem. Exp. Med. 22:303.

Jones, G. M., R. E. Larsen, A. H. Javed, E. Donefer, and J. M. Gaudreau. 1972. Voluntary intake and nutrient digestibility of forages by goats and sheep. J. Anim. Sci. 34:830.

Knight, J. 1965. Some observations on the feeding habits of goats in the South Baringo District of Kenya. E. Afr. Agric. For. J. 30:182.

Kostov, K. 1972. Skorost na razrushavaneto v turbukha na zelena masa ot raigras, chervena vlasatka i ezhova glavitsa. (Rate of breakdown in the rumen of the green matter of ryegrass, red fescue and orchard grass.) Zhivotnovud. Nauki 9:105.

Kronacher, C., J. Kliesch, and H. Schubert. 1934. Anbau- und Fuetterungsversuche mit der gelben Suesslupine. I. Mitteilung: Gruenfuetterungsversuche mit Suesslupine an Schweinen, Ziegen, Schafen und Kuehen. (Agronomic and feeding studies with yellow sweet lupines using swine, goats, sheep and cows.) Z. Tierz. Zuechtgsbiol. Reihe B 21:1.

Lander, P. E., and L. C. Dharmani. 1936. Some digestibility trials on Indian feeding stuffs. X. Green fodders, hays and gram bhusa (chick pea straw). Indian J. Vet. Sci. 6:117.

Larsson, S. 1954. On the hypothalamic organisation of the nervous mechanism regulating food intake. 1. Hyperphagia from stimulation of the hypothalamus and medulla in sheep and goats. Acta Physiol. Scand. 12:1.

Lee, J. W., D. H. Bae, Y. Y. Park, K. S. Kim, and S. H. Park. 1974. Comparative study on digestibility of nutrients in roughages, concentrates and oil meal by Korean native cattle, native goat, sheep, rabbit, pig and poultry. Res. Rep. Off. Rural Dev. (Korea) 16:91.

Louca, A., and A. Papas. 1973. The effect of different proportions of carob pod meal in the diet on the performance of calves and goats. Anim. Prod. 17:139.

Maher, C. 1945. The goat: Friend or foe? E. Afr. Agric. J. 11:115.

Maheshwari, M. L., and S. K. Talapatra. 1975a. Stall feeding of Jamunapari goats with cowpea fodders. Indian Vet. J. 52:30.

Maheshwari, M. L., and S. K. Talapatra. 1975b. Berseem (Egypt. clover) as a fodder for milch goats. Indian Vet. J. 52:113.

Malechek, J. C., and C. L. Leinweber. 1972a. Chemical composition and in vitro digestibility of forage consumed by goats on lightly and heavily stocked ranges. J. Anim. Sci. 35:1014.

Malechek, J. C., and C. L. Leinweber. 1972b. Forage selectivity by goats on lightly and heavily grazed ranges. J. Range Manage. 25:105.

Marten, G. C., and R. N. Andersen. 1975. Forage nutritive value and palatability of 12 common annual weeds. Crop Sci. 15:821.

Martin, F. H., and C. A. Baile. 1972. Feed intake of goats and sheep following acetate or propionate injections into rumen, ruminal pouches, and abomasum as affected by local anesthetics. J. Dairy Sci. 55:606.

McLeod, M. N. 1974. Plant tannins—their role in forage quality. Nutr. Abstr. Rev. 44:803.

McMahan, C. A. 1964. Comparative food habits of deer and three classes of livestock. J. Wildl. Manage. 28:798.

Mehren, M. J., and D. C. Church. 1976. Influence of taste-modifiers on taste responses of pygmy goats. Anim. Prod. 22:255.

Mellenberger, R. W., L. D. Satter, M. A. Millett, and A. J. Baker. 1971. Digestion of aspen, alkali-treated aspen, and aspen bark by goats. J. Anim. Sci. 32:756.

McCamman-Feldman, B. 1980. A critical analysis of tropical savanna forage consumption and utilization by goats. Ph.D. thesis. University of Illinois, Urbana-Champaign.

Merrill, L. B. 1975. Advantages of mixed stocking on rangelands. Beef Cattle Sci. Handb. 12:368.

Merrill, L. B., P. O. Reardon, and C. L. Leinweber. 1966. Cattle, sheep and goats . . . Mix'em up for higher gains. Tex. Agric. Progr. 12:13.

Merrill, L. B., and V. A. Young. 1954. Results of grazing single classes of livestock in combination with several classes when stocking rates are constant. Tex. Agric. Exp. Stn. Progr. Rep. 1726.

Morgen, A. 1912. Ueber die speziellen Wirkungen der Futtermittel auf die Milchproduktion. (Specific effects of feeds upon milk production.) Landwirtsch. Vers. Stn. 77:17.

Morgen, A., C. Beger, and F. Westhausser. 1913. Ueber den Futterwert der eingesaeuerten Zuckerruebenblaetter fuer Milchtiere. (Feeding value of sugar beet leaf silage for milking animals.) Landwirtsch. Vers. Stn. 80:637.

Morgen, A., H. Windheuser, O. Schoeler, and E. Ohlmer. 1922. Ueber die Verdaulichkeit verschiedener Fabrikate entbitterter Lupine und ueber die Verwertung derselben bei der Milchproduktion. (Digestibility of different preparations of nonbitter lupines and utilization for milk production.) Landwirtsch. Vers. Stn. 99:295.

Nastis, A. S. 1977. Consumption, digestion, and utilization by goats of the dry matter and nitrogen in diets containing oak *(Quercus gambelii)* foliage. M.S. thesis. Utah State University, Logan.

Neuhaus, U. 1955. Giftpflanzen fuer Ziegen. (Poisonous plants for goats.) Der Ziegenzuechter 21:1.

Neuhaus, U. 1956. Der Naehrstoffgehalt von frischem Laub, Laubheu und Laubreisig. (Nutritive value of green leaves, dry leaves and twigs.) Dtsch. Kleintierz. 15:2.

Oates, A. V. 1956. Goats as a possible weapon in the control of the thorn bush. Rhodesia Agric. J. 53:68.

Oehmichen, P. 1934. Welche Graeser und Kleearten werden von unseren Ziegen am liebsten gefressen? (Which grasses and legumes are preferred by goats?) Der Ziegenzuechter 29:103.

Oh, H. K., M. B. Jones, and W. M. Longhurst. 1968. Comparison of rumen microbial inhibition resulting from various essential oils isolated from relatively unpalatable plant species. Appl. Microbiol. 16:39.

Poncet, C. 1972. Utilisation digestive comparée de l'orge, de la banane verte et de la banane ensilée chez la chèvre. (Comparative digestion of barley, green plantain and ensiled plantain by goats.) Stn. Rech. l'Elevage Ruminants, Clermont Ferrand, 776.

Provenza, F. D. 1978. Getting the most out of blackbrush. Utah Sci.

Rector, B. S., and J. E. Huston. 1976. Nutrient composition of some Edwards Plateau forage plants. Tex. Agric. Exp. Stn. Bull. 3399.

Sachdeva, K. K., O. P. S. Sengar, S. N. Singh, and I. L. Lindahl. 1973. Studies on goats. 1. Effect of plane of nutrition on the reproduction performance of does. J. Agric. Sci. 80:375.

Saraswat, B. L., K. S. Singh, and K. K. Sachdeva. 1974. A study on the chemical composition of Shisham *(Dalbergia sissoo)* pods at different stages of maturity. Balwant Vidyapeeth J. Agric. Sci. Res. 14:12.

Saxena, J. S. 1971. Studies on Jamunapari goats. I. The chemical composition of the herbage upon which Jamunapari goats thrive in their home tract. Indian Vet. J. 48:77.

Saxena, J. S., and M. L. Maheshwari. 1971. Studies on Jamunapari goats. 2. Studies on comparative intake of nutrients by browsing goats. Indian Vet. J. 48:173.

Saxena, S. K., D. E. Otterby, J. D. Donker, and A. L. Good. 1972. Effect of alkali treatment of oat straw and two levels of soybean meal on ration utilization and faecal composition. Indian Res. J. 6:112.

Schneider, B. H. 1947. Feeds of the world. Their digestibility and composition. W. Va. Agric. Exp. Stn., Morgantown.

Sharma, V. V., and P. C. Murdia. 1974. Utilization of berseem hay by ruminants. J. Agric. Sci., Cambridge 83:289.

Sharma, V. V., and N. K. Rajora. 1977. Voluntary intake and nutrient digestibility of low-grade roughage by ruminants. J. Agric. Sci., Cambridge 88:75.

Sherman, P. J. 1977. Tropical forage legumes. FAO Plant Prod. Protect. Ser. 2, Rome, Italy.

Short, H. L., R. M. Blair, and A. E. Epps, Jr. 1973. Estimated digestibility of some southern browse tissues. J. Anim. Sci. 36:792.

Short, H. L., R. M. Blair, and C. A. Segelquist. 1974. Fiber composition and forage digestibility by small ruminants. J. Wildl. Manage. 38:197.

Short, H. L., D. E. Medlin, and A. E. Anderson. 1965. Ruminoreticular characteristics of mule deer. J. Mammal. 46:196.

Short, H. L., and J. C. Reagor. 1970. Cell wall digestibility affects forage value of woody twigs. J. Wildl. Manage. 34:964.

Simiane, M. de, B. Broquat, and L. Huguet. 1975. Les fourrages verts, comment augmenter leur utilisation par la chévre laitière. (Green forages, how to increase their utilization by the dairy goat.) L'elevage 46:81.

Singh, A. P., V. S. Upadhyay, N. C. Verma, and A. Rekib. 1974. Nutritive value of 'M. P. Chari' *(Sorghum bicolor)* at different stages of maturity. Indian J. Anim. Sci. 44:153.

Singh, K., M. L. Punj, A. S. Kochar, and I. S. Bhatia. 1972. *In vitro* studies on the utilization of the carbohydrate polymers of the cell wall of tree leaves by rumen microorganisms. Indian J. Dairy Sci. 25:148.

Singh, R. P., K. K. Sachdeva, and O. P. S. Sengar. 1969. Digestible nutrients of grain *(Cicer orientinum)* bhusa and Chhilka when fed to Barbari goats. Balwant Vidyapeeth J. Agric. Sci. Res. 11.

Skjevdal, T. 1970. Kraftfortilskudd til mjolkegeiter pa fjellbeite. (Feed concentrate supply for milking goats in mountain pasturage.) p. 123 *In* Husdyrforsoksmotet Norges Landbrukshogsk.

Sidahmed, A. E., J. G. Morris, and S. R. Radosevich. 1981a. Summer diet of Spanish goats grazing chaparral. J. Range Manage. 34:33.

Sidahmed, A. E., J. G. Morris, L. J. Koong, and S. R. Radosevich. 1981b. The contribution of 3 chaparral shrubs to the protein and energy requirements of Spanish goats. J. Anim. Sci. (in press).

Staples, R. R., H. E. Hornby, and R. M. Hornby. 1942. A study of the comparative effects of goats and cattle on a mixed grass-brush pasture. E. Afr. Agric. J. 8:62.

Talbot, L. M., and M. H. Talbot. 1963. The high biomass of wild ungulates on East African savanna. North Am. Wildl. Nat. Res. Conf. Proc. 28:465.

Tessema, S. 1972. Nutritional value of some tropical grass species as compared to some temperate grass species. Ph.D. thesis. Cornell University, Ithaca, New York.

Tetteh, A. 1974. Preliminary observations on preference of herbage species by cattle, sheep and goats grazing on range on the Achimota Experimental Farm. Ghana J. Agric. Sci. 7:191.

Thomas, B., and A. Smith. 1954. The nutritive value of *Calluna vulgaris* (Ling heather). J. Agric. Sci., Cambridge 45:1.

Tripathi, A. K. 1974. Effect of banana leaves tannin and tannic acid on protein degradation. Indian J. Dairy Sci. 27:179.

Upadhyay, V. S., A. Rekib, and P. S. Pathak. 1974. Nutritive value of *Leucaena leucocephala* (Lam.). De Wit Indian Vet. J. 51:534.

Varela, G., J. Ferrer, and J. Boza. 1961. Digestibility trials with goats. Proc. 8th Int. Congr. Anim. Prod., Hamburg, W. Germany.

Verma, N. C., and R. Mukherjee. 1972. Ensilage of berseem *(Trifolium alexandrinum)*. Indian J. Dairy Sci. 25:295.

Van Soest, P. J. 1980. Impact of feeding behavior and digestive capacity on nutritional response. Paper presented at the Technical Consultation on Animal Genetics Resources Conservation and Management, Rome, Italy, June 2-6.

Welch, J. G., D. B. Clark, and J. J. Rutledge. 1975. Body size and rumination in cattle, sheep and goats. J. Anim. Sci. 41:432.

Wilson, A. D. 1977. The digestibility and voluntary intake of the leaves of trees and shrubs of sheep and goats. Aust. J. Agric. Res. 28:501.

Wilson, A. D., J. H. Leigh, N. L. Hindley, and W. E. Mulham. 1975. Comparison of the diets of goats and sheep on a *Casaurina cristata-Heterodendrum oleifolium* woodland community in Western New South Wales. Aust. J. Exp. Agric. Anim. Husb. 15:45.

Wilson, A. D., W. C. Wier, and D. T. Torell. 1971. Comparison of methods of estimating the digestibility of range forage and browse. J. Anim. Sci. 32:1046.

Wilson, P. N. 1957. Studies of the browsing and reproductive behavior of the East African dwarf goat. E. Afr. Agric. J. 23:138.

Wilson, P. N. 1958a. The effect of plane of nutrition on the growth and development of the East African dwarf goat. I. Effect of plane of nutrition on the liveweight gains and the external measurements of kids. J. Agric. Sci. 50:198.

Wilson, P. N. 1958b. The effect of plane of nutrition on the growth and development of the East African dwarf goat. II. Age changes in the carcass composition of female kids. J. Agric. Sci. 51:4.

Wilson, P. N. 1960. The effect of plane of nutrition on the growth and development of the East African dwarf goat. III. The effect of plane of nutrition and sex on the carcass composition of the kid at two stages of growth, 16 lbs. weight, and 30 lbs. weight. J. Agric. Sci. 54:105.

Woods, C. D. 1903. Notes on the Angora goat. Maine Agric. Exp. Stn. Bull. 98.

Zannelli, C. 1958. Influenza dell'alimentazione mista ed esclusiva con sottoprodotti dell'industria agrumaria sui costituenti del sangue della capra. 1. Influenza sul quadro ematologico. (Influence of mixed feeding and feeding only on by-products of the citrus industry on blood constituents of the goat. 1. Influence on the blood picture.) Boll. Soc. Ital. Biol. Sper. 34:44.

RATION FORMULATION AND EXAMPLES OF TYPICAL RATIONS

Agricultural Research Council. 1965. The nutrient requirements of farm livestock. 2. Ruminants. Technical reviews and summaries. Agric. Res. Counc., London, England 11:264.

Beger, C. 1907. Untersuchungen ueber die Einwirkung von Nahrungsfett als Emulsion und als Substanz auf die Milchproduktion. (Effects of feed fat as emulsion and normal upon milk production.) Landwirtsch. Vers. Stn. 67:1.

Blaxter, K. L. 1965. The nutrient requirement of farm livestock. 2. Ruminants. Agric. Res. Counc., London, England.

Cook, L. J., T. W. Scott, S. C. Mills, A. C. Fogerty, and A. R. Johnson. 1976. Effects of protected cyclopropene fatty acids on the composition of ruminant milk fat (lactating cows and goats) diets. Lipids 11:705.

Cunningham, H. M., and J. K. Loosli. 1954. The effect of fat-free diets on lambs and goats. J. Anim. Sci. 13:265.

DeHaas, H. J. 1975. Schafe und Ziegen zur Fleisch-Erzeugung am tropischen Standort. (Sheep and goats for meat production under tropical conditions.) Tropenlandwirt 76:48.

Devendra, C. 1980. Feeding and nutrition of goats, p. 240. *In* D. C. Church, ed., Digestive physiology and nutrition of ruminants. Oregon State University, Corvallis.

Dolezel, F., S. Bartos, and A. Jukl. 1973. Vliv stoupajicich davek sacharozy na produkci tekavych mastnych kyselin v bachoru. (Effect of increasing amounts of sucrose on production of volatile fatty acids in the rumen.) Zivocisna Vyroba 18:715.

Fehr, P. M. 1971. Le chevreau de boucherie. (Production of small goats for slaughter.) Rev. Elevage 26:121.

Fehr, P. M. 1976. Caprins. (Goats.) Tech. Agric. 6:3500. Inst. Techn. d'Elevage Orin et Caprin, Paris, France.

Fehr, P. M., and R. Disset. 1969. Peut-on alimenter rationnellement les chèvres? (Is it possible to feed goats rationally?) Rev. Elevage Special 45:145.

Fehr, P. M., and R. Disset. 1972. Alimentazione delle caprette. (Feeding of young goats.) Inf. Zootec. 19:16.

Fehr, P. M., and D. Sauvant. 1975. L'alimentation avant et après la mise bas: Clé du succes en clevage caprin. (Feeding before and after kidding: The key to success in the rearing of goats.) Elevage 37:83.

Fehr, P. M., and D. Sauvant. 1978. Nutrition and optimum performance of dairy goats. Livest. Prod. Sci. 5:203.

Fehr, P. M., D. Sauvant, J. Delage, B. L. Dumont, and G. Roy. 1976. Effect of feeding methods and age at slaughter on growth performances and carcass characteristics of entire young male goats. Livest. Prod. Sci. 3:183.

Fehr, P. M., and M. deSimiane. 1974. Alimentation des jeunes caprins: L'allaitement artificiel des chevrettes. (Feeding young goats: Artificial nursing of kids.) Bull. Tech. Inf. Minist. Agric. 290:425.

Fingerling, G. 1906. Einfluss fettreicher und fettarmer Kraftfuttermittel auf die Milchsekretion bei verschiedenem Grundfutter. (Effects of fat-rich and fat-poor concentrates on milk secretion with different basic feeds.) Landwirtsch. Vers. Stn. 64:299.

Fowle, K. E., and D. C. Church. 1973. Effect of nutritionally poor ration, short-term fasting, and refeeding on goats. Am. J. Vet. Res. 34:849.

Gaines, W. L. 1943. Feeding standard equations for cows and goats in milk. J. Anim. Sci. 2:304.

Geoffroy, F. 1974. Etude comparée du comportement alimentaire et merycique de deux petits ruminants: La chèvre et le mouton. (Comparison of feeding and ruminating behavior in two small ruminants: Goats and sheep.) Ann. Zootech. 23:63.

Gieseke, A. 1930. Fuetterungsversuch mit Oelkuchenmischfutter bei Ziegen. (Feeding experiment with oil meals for goats.) Mitt. Landwirtsch. 45:955.

Gilch, O. 1930. Ziegenfuetterung auf wissenschaftlicher Grundlage. (Goat feeding scientifically.) Sueddtsch. Landwirtsch. Tierz. 24:65.

Grassland Research Institute. 1964. Roughages for goats in lactation. Annu. Exp. Prog. Rep. 16, Hurley, England.

Haenlein, G. F. W. 1978a. Dairy goat management. J. Dairy Sci. 61:1011.

Haenlein, G. F. W. 1978b. Dairy goats do well on free choice feeding. Hoard's Dairyman 123:1194.

Hino, T., and M. Kametaka. 1974. Effect of diets on the number of protozoa in the rumen, with special reference to the effect of purified diets and sterol in diets. Jpn. J. Zootech. Sci. 45:223.

Holmes, W., G. W. Arnold, and A. L. Provan. 1960. Bulk feeds for milk-production. I. The influence of level of concentrate feeding in addition to silage and hay on milk yield and milk composition. J. Dairy Res. 27:191.

Huston, J. E. 1976. Considerations in the nutrition of milking goats. Proc. 1st Annu. Nev. Dairy Goat Day, Reno, Nevada.

Hutjens, M. F. 1979. Feeding the dairy goat—challenges and opportunities. Dairy Goat J. 57:16.

Hutjens, M. F., and L. H. Schultz. 1971. Effect of feeding soybeans or formaldehyde-treated soybeans on lipid metabolism in ruminants. J. Dairy Sci. 54:1876.

Itabashi, H., and M. Kandatsu. 1974. Effect of rumen ciliate protozoa on the digestibility of food constituents, nitrogen retention and amino acid compositions of the rumen contents. Jpn. J. Zoo. Sci. 45:585.

Jahn, S. 1957. Verwendung von Amidschnitzeln bei Wiederkaeuern nach Versuchen an laktierenden und graviden Milchziegen. (Utilization of amid beet pulp by ruminants after experiments with milking and pregnant goats.) Kuehn-Arch. 71:75.

Johri, C. B., and S. K. Talapatra. 1971. Growth studies with Jamunapari goats. 3. Goat as a meat producing animal. Indian Vet. J. 48:602.

Kanus, S., K. Dogan, and M. Erdem. 1974. Ispirto mayasi'nin oglaklarin canli agirlik artisi, yem tuketimi ve yemi degerlendirme sayisina etkisi uzerinde arastirmalar. (Effect of distillers' dried solubles on liveweight gain, feed consumption and efficiency of utilization of feed by kids.) Uretme, Turkey 5:32.

Kennett, W. S., and L. S. Bull. 1972. The use of feather meal in ruminant diets. Proc. Md. Univ. Nutr. Conf.

Kliesch, J. 1936. Neuzeitliche Fuetterungsversuche an Ziegen. (Modern feeding studies with goats.) Forschungsdienst 1:758.

Komaki, H., K. Osafune, and H. Morimoto. 1974. Study of polyethylene-rope as roughage substitute ration for ruminants. I. Effect on digestibility and VFA (volatile fatty acid) in the rumen digesta of the goat. Bull. Coll. Agric. Vet. Med. Nihon Univ. 31:336.

Kronacher, C., and J. Kliesch. 1928. Die Koerperentwicklung der Ziege von der Geburt bis zum Alter von einem Jahr unter Beruecksichtigung des Naehrstoffbedarfs und der Naehrstoffverwertung der Laemmer sowie der Ernaehrung und Leistungen der Muttertiers. (Body development of goats including nutrient requirements and utilization by kids and the feeding and production of does.) Z. Tierz. Zuechtungsbiol. 11:150.

Lezius, G. 1953. Naehrstoffbedarf und Naehrstoffhaushalt der Milchziege. (Nutrient requirements and balances of dairy goats.) Ph.D. thesis. University of Hohenheim-Stuttgart, West Germany.

Lindahl, I. L. 1972. Nutrition and feeding of goats. *In* D. C. Church, ed., Digestive physiology and nutrition of ruminants. Oregon State University, Corvallis.

Lozano, J. A., and E. Vazquez Salinas. 1972. Algunos efectos de la suplementacion con sorgo antes y despues del parto de cabras estabuladas. (Some effects of sorghum supplementation before and after parturition of goats in stables.) Tierra (Mexico) 27:747.

Machens, A. 1930. Fuetterung, Haltung, Pflege der Ziege. (Feeding, management and care of goats.) Anleitungen Dtsch. Ges. Zuechtungskunde Heft 7.

Machens, A. 1933. Fuetterungsversuche mit Milchziegenleistungsfutter "Ibeka" und Leistungsfutter "Caprina." (Feeding studies with special supplements "Ibeka" and "Caprina.") Der Ziegenzuechter 28:1.

Mead, S. W., O. Peck, and H. H. Cole. 1955. Feed mixtures for milk goats, feeding trials with milking does compared simple and complex concentrate mixtures as the supplements to roughage. Calif. Agric. 9:11.

Morgen, A., C. Beger, and F. Westhausser. 1911. Ueber den Einfluss der aus Futterrueben und Trockenschnitzeln hergestellten Extrakte auf den Fettgehalt der Milch. (Effects of extracts from fodder beets and dry beet pulp upon fat content of milk.) Landwirtsch. Vers. Stn. 75:349.

Morgen, A., W. Wagner, O. Schoeler, and H. Ohlmer. 1918. Fuetterungsversuche mit Leimkraftfutter an Milchtieren. (Feeding experiments with glue by-products on milking animals.) Landwirtsch. Vers. Stn. 94:219.

National Research Council. 1973 and 1978. The nutrient requirements of horses. 3rd and 4th rev. ed. National Academy of Sciences, Washington, D.C.

Needham, K. 1959. Feeding the milking goat. J. Agric. W. Aust. 8:102.

Nitter, G. 1975. Vorlaeufige Ergebnisse mit der Intensivmast von Ziegenlaemmern. (Preliminary fattening results with goat kids.) Kleinviehzuechter 23:316.

Opstvedt, J. 1963. Yields from goats on mountain pasture. Sau og Geit 16:57.

Opstvedt, J. 1967. Feeding experiments with goats. I. Studies on the effect of energy level and feed combination on feed intake and milk yield, composition and organoleptic properties. Tech. Bull. Inst. Anim. Nutr. Agric. Coll. Norway 134:114.

Opstvedt, J. 1969. Norwegian experiments on nutrition and milk quality in goats. Inst. Husdyrernaering Foringslaere Saertr. 343:14.

Rem, S. 1949. Verfuetterung von Sojabohnen. (Feeding of soybeans.) Der Ziegenzuechter 40:169.

Renner, V. 1922. Naehrstoffbedarf und Milchleistung des Rindes und der Ziege. (Nutrient requirements and milk production of cows and goats.) Fruehl. Landwirtsch. Z. 71:335.

Riemenschneider, R. W., and N. R. Ellis. 1936. The effect of ingested cottonseed meal upon the distribution of the constituent fatty acids of goat milk. J. Biol. Chem. 114:441.

Ritgen, G. 1926. Untersuchungen ueber Leistung, Fuetterung, Wachstum rheinischer Ziegen. (Studies on production, feeding and growth of Rhineland goats.) Ph.D. thesis. Agricultural College, Bonn, West Germany.

Robstad, A. M. 1975. Ulik slattetid ved ensilering i relasjon til foring av mjolkegeiter. (Different times for haying in relation to ensilaging in order to produce feed for dairy goats.) Aktuelt Landbruksdep Opplysningstjeneste 1:203.

Sachdeva, K. K., O. P. S. Sengar, S. N. Singh, and I. L. Lindahl. 1973. Studies on goats. 1. Effect of plane of nutrition on the reproductive performance of does. J. Agric. Sci., Cambridge 80:375.

Sachdeva, K. K., O. P. S. Sengar, S. N. Singh, and I. L. Lindahl. 1974. Studies on goats. 2. Effect of plane of nutrition on milk production and composition. Milchw. 29:471.

Schmidt, J., and H. Vogel. 1932. Ziegenfuetterungsversuche mit Herdbuchfutter B und DLG Mischfutter II b. (Goat feeding experiments with herd feed B and DLG feed II b.) Der Ziegenzuechter 33:257.

Schmidt, J., H. Vogel, and I. Duckstein. 1932. Vergleichende fuetterungsversuche mit Ibeka-Leistungsfutter und DLG Mischfutter IIb an Ziegen. (Comparative feeding experiments with Ibeka and DLG feed IIb for goats.) Sueddtsch. Landwirtsch. Tierz. 26:209.

Schmidt, J., H. Vogel, and I. Duckstein. 1933. Ziegenfuetterungsversuch mit DLG. Milschfutter IIb bei wechselnder Futternorm. (Goat feeding experiments with DLG feed IIb under different standards.) Mitt. Dtsch. Landwirtsch. Ges. 17:360.

Schmidt, J., H. Vogel, and I. Duckstein. 1934. Vergleichender Ziegenfuetterungsversuch mit verschiedenen Gruenfutterpflanzen. (Comparative goat feeding experiments with different green plants.) Dtsch. Landwirtsch. Tierz. 38:21.

Skjevdal, T. 1972. Supplements of concentrates for dairy goats on mountain pasture. Nor. Landbrukshogsk. 401:190.

Skjevdal, T. 1972. Comparison between silage and silage combined with hay as a roughage for milk goats. Sau og Geit. 25:23.

Skjevdal, T. 1973. Kraftfortilskudd til geiter pa beite. (Feed supplements for goats on pasture.) Landbrukstidende 79:551.

Skjevdal, T. 1974. Potatoes and swedes in the diet of ruminants. III. Studies in lactating goats. Meld. Nor. Landbrukshogsk. 169.

Skjevdal, T. 1974. Milk feeding of kids. Meld. Nor. Landbrukshogsk. 53:39.

Skjevdal, T. 1979. Effect of feeding various levels of concentrate during the dry period in dairy goats. Meld. Nor. Landbrukshogsk. 58:24.

Sutoh, H., S. Uchida, and K. Miyake. 1972. Studies on silagemaking. 19. The relations between the growth stage and silage quality of the green oats. Sci. Rep. Okayama Univ. 40:25.

Tanaka, K., I. Nakajima, and H. Hayashi. 1973. Effect of safflower oil supplement on the plasma lipids of goats. Jpn. J. Zool. Sci. 44:19.

Tasaki, I. 1951. Studies on the metabolism of dairy goats. I. Metabolism in the fasting condition. Jpn. J. Zootech. Sci. 25:1.

Uchida, S., H. Sutoh, and S. Imai. 1972. Studies on silagemaking. 20. Chemical composition of rice plants at different growth stages and feeding value of rice plant silage and rice plant hay. Sci. Rep. Okayama Univ. 40:35.

Ulvesli, O., M. Husby, and K. Presthegge. 1952. Forkontroll med geiter og undersokelser av geitemelk. (Feed consumption of goats and studies of goats' milk.) Tidsskr. Nor. Landbruk 59:118.

Williams, H. H., and L. A. Maynard. 1934. The effect of specific dietary fats on the blood lipids of lactating goats. J. Dairy Sci. 17:223.

Wilson, P. N. 1957. Effect of contrasted planes of nutrition on the carcass composition of the East African dwarf goat. Nature 180:145.

Winter, J., and R. Goersch. 1974. Ziegen als Versuchstiere—ein Beitrag zur Fuetterungsoptimierung. (Goats as experimental animals: A contribution to feeding optimization.) Z. Versuchstierkd. 16:256.

Wolter, R. 1971. Rationnement pratique de la vache laitière, de la chèvre et des ovins. (Practical nutrition of the milking cow, the goat, and the sheep.) Vigot Freres 112.

Zwagermann, C. 1921. Das Fuettern der Ziegen und Futternormen. (The feeding of goats and feeding standards.) Der Ziegenzuechter 16:362.

NUTRITION-RELATED METABOLIC DISORDERS

Adam, S. E. I., G. Tartour, H. M. Obeid, and O. F. Idris. 1973. Effects of *Ipomoea carnea* on the liver and on serum enzymes in young ruminants. J. Comp. Path. 83:531.

Alton, G. G. 1973. Brucellosis in goats and sheep. World Anim. Rev. 5:16.

Anderson, J. J. B. 1970. Parturient hypocalcemia. Academic Press, New York.

Anderson, J. J. B., Y. H. Kuo, W. C. Crackel, and A. R. Twardock. 1970. Effects of high and low dietary calcium on the size and half life of the readily exchangeable calcium pool in female goats under two different feeding conditions, p. 147. *In* J. J. B. Anderson, Parturient hypocalcemia. Academic Press, New York.

Barajas Rojas, J. A., and E. L. Biberstein. 1974. The dipteroid agent of ovine posthitis: Its relationship to *Corynebacterium renale*. J. Comp. Path. 84:301.

Binns, W., R. F. Keeler, and L. D. Balls. 1972. Congenital deformities in lambs, calves, and goats resulting from maternal ingestion of *Veratrum californicum:* hare lip, cleft palate, ataxia, and hypoplasia of metacarpal and metatarsal bones. Clin. Toxicol. 5:245.

Crookshank, H. R. 1966. Effect of sodium or potassium on ovine urinary calculi. J. Anim. Sci. 25:1005.

Crookshank, H. R. 1970. Effect of ammonium salts on the production of ovine urinary calculi. J. Anim. Sci. 30:1002.

Dollahite, J. W. 1961. Shin oak *(Quercus havardii)* poisoning in cattle. Southwest. Vet. 14:198.

Dollahite, J. W. 1972. The use of sheep and goats to control *Senecio* poisoning in cattle. Southwest. Vet. 25:223.

Griffith, D. P., D. M. Musher, and C. Itin. 1975. Urease—the primary cause of infective—induced urinary stones. Investig. Urol. 13:346.

Guessous, F., P. M. Fehr, and J. Delage. 1974. Influence de l'acetate et du propionate de Na ajoutés au régime avant et après la mise bas sur le métabolisme, la production et la composition lipidique du lait de chèvre. (Effect of sodium acetate and propionate added to the diet before and after parturition on metabolism, production and lipid composition of the milk of goats.) Ann. Biol. Anim. Biochem. Biophys. 14:251.

Gunji, S., K. Shinozaki, and A. Shiga. 1974. Studies on rumen fermentation in ruminants. II. Effects of several chemical additives introduced into rumen on rumen fermentation and blood glucose in goats. J. Fac. Agric. Iwate Univ. 11:281.

Guss, S. 1977. Management and diseases of dairy goats. Dairy Goat J. Publ., Scottsdale, Ariz.

Hardy, W. T., V. L. Cory, H. Schmidt, and W. H. Dameron. 1931. Bitterweed poisoning in sheep. Tex. Agric. Exp. Stn. Bull. 433.

Hoar, D. W., R. J. Emerick, and L. B. Embry. 1970. Potassium, phosphorus and calcium interrelationships influencing feedlot performance and phosphatic urolithiasis in lambs. J. Anim. Sci. 30:597.

Hume, I. D., R. J. Moir, and M. Somers. 1970. Synthesis of microbial protein in the rumen. I. Influence of the level of nitrogen intake. Aust. J. Agric. Res. 21:283.

Idris, O. F., G. Tartour, S. E. I. Adam, and H. M. Obeid. 1973. Toxicity to goats of *Ipomoea carnea*. Trop. Anim. Health Prod. 5:119.

James, C. S., and K. Chandran. Enquiry into the role of minerals in experimental urolithiasis in goats. Indian Vet. J. 52:251.

Keeler, R. F., K. R. Van Kampen, and L. F. James. 1978. Effects of poisonous plants on livestock. Proc. Joint U.S.-Aust. Symp. Poisonous Plants, Academic Press, New York.

Kingsbury, J. M. 1964. Poisonous plants of the United States and Canada. Prentice-Hall, Englewood Cliffs, N.J.

Kromann, R. P., A. E. Joyner, and J. E. Sharp. 1971. Influence of certain nutritional and physiological factors on urea toxicity in sheep. J. Anim. Sci. 32:732.

Ku Suan Zan. 1973. Content of volatile fatty acids in the rumen and abomasum of goats continuously given sodium propionate in the feed. Zhivotnovud. Nauki 10:35.

Leach, C. E. 1971. The goat owners scrap book. American Supply House, Columbia, Missouri.

Lindahl, I. L. 1972. Nutrition and feeding of goats. *In* D. C. Church, ed., Digestive physiology and nutrition of ruminants. Oregon State University, Corvallis.

Littledike, E. T. 1974. Parturient hypocalcemia, hypomagnesemia, mastitis-metritis-agalactia complex of swine, p. 355. *In* B. L. Larson and V. R. Smith, Lactation, vol. II. Academic Press, New York.

Mackenzie, D. 1973. Goat husbandry. Faber & Faber, London, England.

Majumdar, B., and B. N. Gupta. 1959. Vitamin A deficiency and urinary calculi in goats. Ann. Biochem. Exp. Med. 19:249.

Mathews, F. P. 1936. The toxicity of broomwood *(Gutierreza microcephala)* for sheep, and cattle and goats. J. Am. Vet. Med. Assoc. 88:56.

McMillan, O., and B. Southcott. 1973. Aethiological factors in ovine posthitis. Aust. Vet. J. 49:405.

Meites, J., H. D. Webster, F. W. Young, F. Thorp, Jr., and R. N. Hatch. 1951. Effects of corpora lutea removal and replacement with progesterone on pregnancy in goats. J. Anim. Sci. 10:411.

Menahan, L. A. 1966. Interrelationships between ketone body production, carbohydrate utilization and fat mobilization in the ruminant. Ph.D. thesis. University of Wisconsin, Madison.

Morris, J. G., and E. Payne. 1970. Ammonia and urea toxicoses in sheep and their relation to dietary nitrogen intake. J. Agric. Sci. 74:259.

Muir, L. A., E. L. Rickes, P. F. Duquette, and G. E. Smith. 1981. Prevention of induced lactic acidosis in cattle by thiopeptin. J. Anim. Sci. 52:635.

Nichol, O. 1937. Vitamin robbery by parasitic worms. Br. J. Nutr. 6:47.

Omori, S., and H. Sato. 1973. Some aspects of clinical signs and blood chemical composition of urea poisoning in the ruminants. Bull. Natl. Inst. Anim. India 26:9.

Osborne, W. B., and F. A. Widdows. 1961. Sheath rot and testosterone—its effects and side effects on health and production. Wool Technol. Sheep Breed. 8:99.

Radloff, H. D. 1964. Factors affecting plasma free fatty acid concentrations in the ruminant. Ph.D. thesis. University of Wisconsin, Madison.

Reynolds, M. 1970. Mammary circulatory and respiratory changes associated with parturition in goats, p. 97. *In* J. J. B. Anderson, Parturient hypocalcemia. Academic Press, New York.

Robbins, G. D., H. O. Kunkel, and H. R. Crookshank. 1965. Relationship of dietary mineral intake to urinary mineral excretion and incidence of urinary calculi in lambs. J. Anim. Sci. 24:76.

Sato, H., and S. Omori. 1977. Incidence of urinary calculi in goats fed a high phosphorus diet. Jpn. J. Vet. Sci. 39:531.

Schultze, A. B., and C. W. Turner. 1948. Relation of endocrine gland weight to body weight in growing and mature female dairy goats. Missouri Agric. Exp. Stn. Res. Bull. 427.

Schultz, L. H. 1974. Ketosis. *In* B. L. Larson and V. R. Smith, Lactation, vol. 1. Academic Press, New York.

Schwalm, J. W., and L. H. Schultz. 1974. Changes in blood and liver metabolites in experimental diabetes in fed and fasted goats. J. Dairy Sci. 57:606.

Schwalm, J. W., and L. H. Schultz. 1976. Blood and liver metabolites in fed and fasted diabetic goats. J. Dairy Sci. 59:262.

Shelton, M., and W. C. Ellis. 1965. Buffering agents in all concentrate rations for lambs. J. Anim. Sci. 24:289.

Shelton, M., and J. Groff. 1974. Reproductive efficiency in Angora goats. Tex. Agric. Exp. Stn. Bull. 1136.

Shelton, M., and C. W. Livingston. 1975. Posthitis in Angora wethers. J. Am. Vet. Med. Assoc. 167:154.

Shelton, M., and J. R. Stewart. 1973. Partitioning losses in reproductive efficiency in Angora goats. Tex. Agric. Exp. Stn. Bull. 3187.

Smith, D. D. 1975. Grazing studies on selected plutonium-contaminated areas in Nevada. *In* M. G. White and P. B. Dunaway, The radioecology of plutonium and other transuranics in desert environments. U.S. ERDA, Las Vegas.

Sperry, O. G., J. W. Dollahite, G. O. Hoffman, and B. J. Camp. 1964. Texas plants poisonous to livestock. Tex. Agric. Exp. Stn. Bull. 1028.

Tartour, G., S. E. I. Adam, H. M. Obeid, and O. F. Idrus. 1974. Development of anemia in goats fed with *Ipomoea carnea*. Br. Vet. J. 130:271.

Twardock, A. R., Y. H. Kuo, W. C. Crackel, and J. J. B. Anderson. 1970. Measurement of calcium kinetics in dairy goats, p. 135. *In* J. J. B. Anderson, Parturient hypocalcemia. Academic Press, New York.

Van Heerden, K. M. 1963. Investigations into the cause of abortions in Angora goats in South Africa. Onderstepoort J. Vet. Res. 30:23.

Van Rensburg, S. J. 1964. Possible endocrine mechanisms for gestation failure. Proc. 5th Int. Congr. Anim. Reprod. Artif. Insem., Trento, Italy.

Van Rensburg, S. J. 1971. Malnutrition of the fetus as a cause of abortion. J. S. Afr. Vet. Med. Assoc. 42:305.

Vercoe, J. E. 1969. Transfer of nitrogen from the blood to the rumen in cattle. Aust. J. Agric. Res. 20:191.

Wentzel, D., M. M. LeRoux, and L. J. J. Botha. 1976. Effect of the level of nutrition on blood glucose concentration and reproductive performance of pregnant Angora goats. Agroanimalia 8:59.

Wentzel, D., J. C. Morgenthal, C. H. Van Niekerk, and C. S. Roelofse. 1974. The habitual aborting Angora doe. II. The effect of an energy deficiency on the incidence of abortion. Agroanimalia 6:129.

Wentzel, D., J. C. Morgenthal, and C. H. Van Niekerk. 1975. The habitually aborting Angora doe. III. Luteal function in normal and aborter does. Agroanimalia 7:15.

Wentzel, D., J. C. Morgenthal, and C. H. Van Niekerk. 1975. The habitually aborting Angora doe. V. Plasma estrogen concentration in normal and aborter does with special reference to the effect of an energy deficiency. Agroanimalia 7:35.

Wentzel, D., and C. S. Roelofse. 1975. The habitually aborting Angora doe. VII. Induction of abortion by administration of cortisone acetate. Agroanimalia 7:45.

Wentzel, D., J. M. van der Westhuysen, J. H. P. van der Merwe, and K. S. Viljoen. 1976. Effect of intrafoetal maternal oestrogen levels and subsequent reproductive outcome of pregnant Angora does. Agroanimalia 8:39.

FEED COMPOSITION

Devendra, C. 1979. Malaysian feedingstuffs. Malays. Agric. Res. Div. Inst., Serdang, Selangor, Malaysia.

Garrett, W. N. 1976. Net energy estimation. Personal Comm. Dept. Anim. Sci., Univ. Calif., Davis.

Gohl, B. I. 1975. Tropical feeds. FAO, Rome, Italy.

Harris, L. E., H. Haendler, R. Riviere, and L. Rechaussat. 1980. International feed data bank system. Int. Feedstuffs Inst. Publ. 2, Utah State University, Logan.

Harris, L. E., L. C. Kearl, and P. V. Fonnesbeck. 1972. Use of regression equations in predicting availability of energy and protein. J. Anim. Sci. 35:658.

Kearl, L. C., L. E. Harris, H. Lloyd, M. F. A. Farid, and M. F. Wardeh. 1979. Arab and Middle East tables of feed composition. Utah State University, Logan.

McDowell, L. R., J. H. Conrad, J. E. Thomas, and L. E. Harris. 1974. Latin American tables of feed composition. University of Florida, Gainesville.

Moe, P. W., and H. F. Tyrrell. 1976. Estimating metabolizable and net energy of feeds. Proc. 1st Int. Symp. Feed Comp., Anim. Nutr. Req. Comput. Diets, Utah State University, Logan.

National Academy of Sciences. 1969. United States–Canadian tables of feed composition. National Academy of Sciences Publ. 1684, Washington, D.C.

National Academy of Sciences. 1971. Atlas of nutritional data on United States and Canadian feeds. National Academy of Sciences Publ. 1919, Washington, D.C.

Sen, K. C., and S. N. Ray. 1971. Nutritive value of Indian cattle feeds and the feeding of animals. Indian Counc. of Agric. Res., New Dehli.